Ancient Loons

Ancient Loons

Stories David Pingree Told Me

Philip J. Davis

CRC Press
Taylor & Francis Group
Boca Raton London New York

CRC Press is an imprint of the
Taylor & Francis Group, an **Informa** business

AN A K PETERS BOOK

CRC Press
Taylor & Francis Group
6000 Broken Sound Parkway NW, Suite 300
Boca Raton, FL 33487-2742

First issued in paperback 2017

ISBN 13: 978-1-138-11453-1 (pbk)
ISBN 13: 978-1-4665-0111-9 (hbk)

This book contains information obtained from authentic and highly regarded sources. Reasonable efforts have been made to publish reliable data and information, but the author and publisher cannot assume responsibility for the validity of all materials or the consequences of their use. The authors and publishers have attempted to trace the copyright holders of all material reproduced in this publication and apologize to copyright holders if permission to publish in this form has not been obtained. If any copyright material has not been acknowledged please write and let us know so we may rectify in any future reprint.

Library of Congress Cataloging-in-Publication Data

Davis, Philip J., 1923-
   Ancient loons : stories David Pingree told me / Philip J. Davis.
      p. cm.
   "An A K Peters book."
   Includes bibliographical references.
   ISBN 978-1-4665-0111-9 (hardback)
   1. Science--Miscellanea. 2. Scientists--Intellectual life--Miscellanea. I. Title.

Q173.D254 2012
509.2'2--dc23   2011035759

Visit the Taylor & Francis Web site at
http://www.taylorandfrancis.com

and the CRC Press Web site at
http://www.crcpress.com

Once again, to Hadassah—and to all the storytellers of the world.

Table of Contents

Preface

This slim volume tells of my relations with David E. Pingree: a classicist, an Orientalist, a historian of ancient mathematics, and for many years a member of the Department of the History of Mathematics at Brown University. I intend here to play Boswell to David's Johnson; no, Watson to Sherlock Holmes; no, even more appropriately, to play Archie to Nero Wolf while I retell some of the stories of weird historic and mythic characters that David introduced me to. In telling these stories, I will weave through them a number of personal experiences that bear on our relationship.

I am aware that the very word "mathematics" sends chills up the spine of many people; mathematics is a complete turnoff. However, I can promise my readers that for the most part they will encounter here no mathematics at all and what little there is can be understood by any student who has been promoted successfully to middle school. And if, on a few occasions, I break this promise, it is only to display visually and linguistically and as pure decoration to my otherwise unadorned prose what some of the "higher" mathematical material looks and sounds like. No comprehension is required.

Note: The little bits of dialogue that I have inserted in various places are, I believe, true to the spirit of my conversations with David (and with a few other people). On the other hand, the letters I have inserted are authentic.

Acknowledgments
(Without whom not)

Isabelle Pingree, Amanda Pingree, Janet Sachs-Toomer, Gerald J. Toomer, Marguerite Dorian, Stephanie Han, Alice and Klaus Peters, and Kara Ebrahim

I love anecdotes . . . (but) if a man is to wait till he weaves anecdotes into a system, we may be long in getting them, and get but a few in comparison of what we may get.

—Samuel Johnson, from Boswell's *A Tour of the Hebrides*

Oh, Egypt, Egypt. Of thy religion nothing will remain but graven words, and only the stones will tell of thy piety.

—Hermes Trismegistus

Part I

Setting the Stage:
The Academic Milieu

I n January of 1963 I was at the University of California in Berkeley at the winter meetings of the various national mathematical societies, where I received the Chauvenet Prize of the Mathematical Association of America. On this occasion I made two friendships that turned out to be of great importance to me professionally. At the award ceremony, which was held in a large theater of the university, I responded with a few remarks that at the time were considered shockers. At least this is what Reuben Hersh, who was in the audience, told me a few years later when we were well into a very fruitful collaboration on the philosophy of mathematics.

What was the tendency of these remarks? Simply put: that mathematics was a product of the human brain, produced and fostered by society at certain historical times. This view was opposed to the popular view held by 95% of mathematicians, that their subject and all its contents pre-existed in a platonic universe of ideas, independently of humans. Reuben, who is far more philosophically informed and inclined than I, found that what I said resonated sympathetically with his own views of mathematics. A number of years passed, and ultimately we were brought together productively by a mutual friend.

The second friendship contracted on this occasion was with Janet Sachs (now Sachs-Toomer). At the time, Janet was working at the *Mathematical Reviews*, then located in Providence, and I was about to move to Providence to accept a position in Brown's Department of Applied Mathematics. Janet was very helpful in getting my family settled in Providence. Through Janet I met her husband, Abe Sachs, distinguished cuneiformist and historian of ancient science, and through Abe I met Otto Neugebauer, the world-renowned historian of ancient Egyptian and Babylonian mathematics and the founder of the unique Department of the History of Mathematics at Brown.

The members of the History of Mathematics Department used to have lunch frequently in the university cafeteria known as the Ivy Room. Otto Neugebauer and Abe Sachs were there, together with distinguished visitors and old students such as Asger Aaboe, Bernie Goldstein, Edward Kennedy, Noel Swerdlow, and "visiting firemen" who appeared off and on. I found these fellows to be interesting company and I suppose the reverse was also true; I would join them in the role of Honorary Member or, rather, Kibitzer to the Department.

The permanent members of the department all referred to each other with the names of animals. Neugebauer, the founder of the department and the doyen of the group, was referred to as "The Elephant," a designation I think he created for himself realizing that in American academic circles the Herr Doktor Professor bit was not appropriate. In all the years I knew him I never once heard him referred to as "Professor," or "Neugebauer," or "Otto," but always "The Elephant." I myself could never bring myself to participate in this kind of zoological nomenclature.

At these lunches I recall discussions as to whose theories were wrong or trivial, the sins of museum directors who would not allow scholars to see the archeological finds they were holding, and other such sensational material. An aura of scandal (ancient and contemporary), skepticism, and gossip wafted through the lunch table talk of these men. I would not have been able to follow highly technical material if it had emerged, but mercifully its emergence was rare and I latched on to and enjoyed the light professional banter and fluff.

Wilbour Hall

T he offices of the Department of the History of Mathematics at Brown were in Wilbour Hall, an old brick building located at the corner of George and Prospect Streets. The "George" is King George III of England: Brown predates the American Revolution. The "Prospect" is clear enough when one realizes that the Brown campus is on a hill that overlooks downtown Providence. But who was Wilbour? Therein lies a tale and I shall give only an abridged version of it here. (For more, see my little book *The Thread*.)

Between the years 1868 and 1871, "Boss" Tweed of Tammany Hall and his cronies managed to steal a reputed $45,000,000 from the city of New York. Figuring an inflation rate of 1:10 at the very least, you can see that this was a very fine piece of cash indeed. One of the beneficiaries of this plunder was a man by the name of Charles Edwin Wilbour. Wilbour was descended from an old Rhode Island family that had settled in Little Compton in the 1690s. He was a bright boy with an interest in languages. He entered Brown with the class of 1854 and won a prize in Greek.

In late 1854, Wilbour set out for New York City. He landed a job on the *New York Herald* as a reporter, and as Horace Greely took a liking to him, he

Wilbour Hall at Brown University

rose rapidly in the world of publishing. Through these connections Wilbour met Boss Tweed.

Tweed took over the *New York Transcript* and installed Wilbour as manager. Simultaneously, Tweed bought out a printing firm called the New York Printing Company and installed Wilbour as its president. The *Transcript* became the official newspaper of the city of New York and all municipal advertising was channeled to it. In 1871, Tweed gave orders to the board of education that all textbook bids from *Harper's Weekly* were to be rejected because *Harper's* was publishing Thomas Nast's anti-Tweed cartoons. He gave further instructions that all textbooks published by Harper & Brothers in use in the school system were to be destroyed and replaced by books published by Wilbour's company. In a period of 30 months the bill for public advertising came to more than $7,100,000. The bulk of this sum found its way into the pockets of the New York Printing Company and its officers.

With Boss Tweed's downfall in 1871, Wilbour and his family packed up hurriedly and sailed for France. And now comes the remarkable part of the story. Having buried one career in America, Wilbour took up a second career: he became a gentleman Egyptologist and a very good one.

Wilbour set up in Paris. He gained entrée to the French literati and became friendly with Victor Hugo. He also met Gaston Maspero, who was one of the leading Egyptologists. His interest in the ancient Egyptian civilization was kindled or rekindled by Maspero and he studied on his own for a while. In 1880, he traveled up the Nile on a government gunboat. In 1886, he did it up brown by buying a houseboat—a *dahabiyeh*—and spending every winter on it.

In the last years of his life, Wilbour sailed up and down the Nile, poked into all the temples, read the inscriptions in hieroglyphic or Greek, and acquired a few antiquities every now and then. In 1899, he made some interesting finds on an island in the First Cataract of the Nile. He died in 1896 and was well respected by all Egyptologists even though he never published anything.

Wilbour's collection of antiquities ultimately found their way into the Brooklyn Museum. His daughter Theodora survived her three siblings and was herself without heirs. In 1931, Theodora remembered her father's Alma Mater, even though he had neither graduated from Brown nor taken any interest in the place while he was alive. In 1947, Theodora died and Brown got three-quarters of a million dollars to support a Department of Egyptology. Their interests being close, the Department of the History of Mathematics shared space in the renamed Wilbour Hall with the Egyptologists.

And now to describe the principal members of the Department of the History of Mathematics at Brown University—alas, the department is no longer in existence; a great pity and a shame!—fallen victim to a university malaise that I call *aggravated provositis.*

Otto Neugebauer

I first met Neugebauer (1899–1990) in the fall of 1963 when I joined the faculty of Brown University. For many years, prior to his return to Princeton's Institute of Advanced Studies in 1984, I ate lunch with him fairly often in the Brown cafeteria known as the Ivy Room. Not infrequently, we were joined by other members of the History of Mathematics Department or their visitors. The conversation was relaxed but lively, scholarly but usually very general, and was terminated when the last person had finished lunch. Neugebauer was not one to twiddle his spoon leisurely in a second cup of coffee.

Neugebauer, born in Graz, a preeminent scholar in the history of ancient science and mathematics, had that soft appearance and low-decibel manner that I associate with Austrians. Inwardly he held firm opinions and prejudices, and would occasionally burst out in anger and irritation using English swear words that I felt were an uncomfortable translation from German language originals. Not unlike Mark Twain, he was a misanthrope; he perceived the human world as consisting largely of fools, knaves, and dupes, and when he was overwhelmed by this perception he took refuge in

Otto E. Neugebauer (Image
Courtesy of Brown University)

his love of animals, which was tender and deep. I recall his telling us (when my wife Hadassah and I visited him at his daughter's summer cabin on Deer Island, Maine) how he had made a pet of a fox that came visiting for food. On Deer Island, walking around dressed comfortably in a pair of overalls he had picked up at Epstein's (a general store in the village), he did not play the great professor.

Neugebauer had his roster of The Greats in his profession, and although his ratings were not as finely tuned as those of the English mathematician G. H. Hardy, anyone who ate lunch with him would find out within a week which of the great names were really great and which were intellectual asses. As regards the past, he thought that Copernicus was overrated—he called him Koppernickel. Kepler was much better, and he loved Arthur Koestler's popularization of Kepler in *The Sleepwalkers*. Claudius Ptolemy was a great hero. As regards contemporaries, he expressed his views candidly. I was occasionally shocked and have no desire to go public with them.

Neugebauer could not abide philosophy; he thought it a great waste of time and rarely discussed it. He would parody Hegel. I would guess, though, that his largely unspoken philosophy of science was that of the logical positivism promulgated by the famous Vienna Circle (*Wienerkreis*) in the '20s and early '30s. Insofar as he had formulated a philosophy of mathematics, it was in considerable opposition to mine. But as regards the history of ancient mathematics and science, I sat at his feet.

Though he perceived the tension between the individual culture and the universal aspect of mathematics and admired the writings of the famous Egyptologist Kurt Sethe on these matters, when the chips were down, according to a splendid biographical memoir written by Noel Swerdlow, "even through years of allowing that mathematics was grounded in culture, he really believed that in a more profound sense it was not."

Neugebauer's attitude (I would not call it philosophy) towards education comforted me. I had always considered myself a poor teacher, particularly with thesis students, and thought that students should be given an initial push and then encouraged (or even required) to go their own way. According to Swerdlow this was also Neugebauer's attitude, both for himself and for his students, and he justified it satirically by asserting that "no one had yet invented a system of education that was capable of ruining everyone."

Born into a Protestant family in Catholic Austria he could not abide religious ritual or theological dogmas, and although he was a considerable student of these matters he parodied them all mercilessly. As a relaxation he would read *The Lives of the Ethiopic Saints*, and he made a dossier of the names of the individual devils that those saints had to contend with.

He was given to playful irony and loved Anatole France's ironic fantasies. I recall his being greatly amused by a passage he had read in an ancient Near Eastern medical text that said crocodile droppings were prescribed for such and such a medical condition. A patient applied them without success and went back to the practitioner with a complaint. The practitioner asked,

"What was the sex of the crocodile?"

"Male, I think."

"Then try a female."

Neugebauer commented on this story with a twinkle in his eyes, "You see, the ancients knew all about hormones."

The work that made Neugebauer famous was the reconstruction, understanding, and interpretation of ancient scientific texts—Egyptian, Babylonia, and Ethiopian. He led us to understand the rich body of mathematics created by the pre-Greek Babylonians, whose general civilization he found very uncomfortable. The text was the thing. Correspondingly, I would guess that his philosophy of history (if he ever expressed it openly) would have been that of Leopold von Ranke (1795–1866): *"Die Vergangenheit, wie sie eigentlich war."* (The past as it really was.) I once asked him to comment on the assertion one often sees in print that the stones in Stonehenge, England, were placed so as to line up in this or that manner with the various celestial bodies. I remarked that the field of "paleoastronomy," as it is called, was burgeoning. His reaction was a short and pointed *"Quatsch."* (Nonsense.)

One lunch I happened to suggest that Ptolemy's model of the solar system, with its cycles and epicycles, was an early and interesting example of curve fitting by means of trigonometric series. This thought came to me through my work on approximation theory, but in point of fact I had heard it much earlier

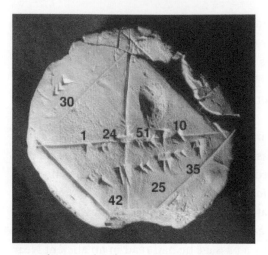

Theorem of Pythagoras in ancient clay

from another man whom I admired greatly: in the lectures of the physicist and philosopher of science Philipp Frank.

Neugebauer bridled at my suggestion. "No! No! No! I will explain to you what Ptolemy was up to." And he got up and left the lunch group. A few days later, I received a two-page handwritten letter from him explaining in detail "what Ptolemy was up to." I was touched that Neugebauer took the time to write to me and felt it was another indication of his total devotion to his historic craft, as well as of his regard for me. This knuckle rapping led to no diminution of our friendship. Our lunch meetings continued, and my wife and I visited him and his daughter several times in their summer home on Deer Island in the Penobscot Bay area.

In one of his prefaces, Neugebauer thanked the ghosts of ancient Babylonian scribes who "by their untiring efforts . . . built the foundations for the understanding of the laws of nature which our generation is applying so successfully to the destruction of civilization." At least, he added, "they also provided hours of peace for those who attempted to decode their lines of thought two thousand years later." I'm sure he would have agreed with Shakespeare—in *Macbeth*—that human history "is a tale told by an idiot, full of sound and fury signifying nothing." But the very separation he was able to make between the human world and the world of platonic ideas was a source of strength and a basis for faith.

Abraham J. Sachs

A be (1915–1984) was a dour, rather silent type, for whom the sunshine occasionally broke through. He was a man totally dedicated to his discipline of Assyriology. Abe had been a student of the famous archaeologist and historian of the ancient Near East: Johns Hopkins Professor William F. Albright (1891–1971), an American Orientalist and Biblical scholar. In 1948, Abe, who was then only a research assistant, was offered the chair of Assyriology at Hopkins, but he preferred to come to Brown to work together with and alongside Neugebauer.

An interpolation: years ago, I once had lunch with Albright at my father-in-law's house. I knew he had made excavations in various places in Palestine, and I asked him (rather foolishly, in retrospect, but I was young then) why anyone should bother digging up the past. Albright didn't answer my question, but gave me a look that I interpreted as "wise-guy kid!"

Abe once told me a personal story that upped my respect for archaeologists and cuneiformists greatly. He once found a cuneiform tablet of

A reassembled cuneiform tablet

considerable interest in one museum. Part of the tablet was broken off but promised to contain material that was vital to the understanding of the whole. Some years later, he was able to find the broken-off part in a totally different museum and the pieces fit together like pieces in a jigsaw puzzle!

For my 60th birthday (60 being of considerable significance in Babylonian numerology), Abe gave me a present of a drawing he had made for me depicting a broken tablet with the number 60 in cuneiform script plus additional comments that he was reluctant to interpret. Abe left behind numerous workbooks that Assyriologists still work with today.

Gerald J. Toomer

In 1965, Gerald Toomer, a brilliant classicist from Oxford, joined the department. Toomer has published fundamental studies of Greek, Arabic, and medieval Latin mathematics and of mathematical astronomy. From his pen has come a commentary on Diocles' *Burning Mirrors* and a translation and annotation of Ptolemy's *Almagest*. His later interests have embraced the study of English Arabists in the 17th century and through this connection he has published *Eastern Wisdom and Learning: The Study of Arabic in Seventeenth Century England* as well as *John Selden: A Life in Scholarship*.

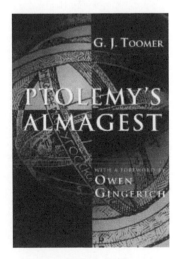

Front cover of Gerald Toomer's translation of the *Almagest*

Enter David Pingree,
My Protagonist

I n the fall of 1971, David Edwin Pingree (1933–2005), who had come from
the University of Chicago, entered this mise en scène—or, more accurately
put since its members referred to one another by animal names, this zoo.
David was a classicist, linguist, Orientalist, historian, avid book collector
and bibliophile, and member of the History of Mathematics Department at
Brown from 1971 to the time of his passing.

David grew up in Providence and went to the Moses Brown School there.
When later his family moved to Andover, Massachusetts, he went to Phillips
Andover Academy. There he taught himself Sanskrit and it was then that he
began to think about the transmission of scientific information across cul-
tures, a topic that engaged him professionally for the rest of his life. From
Phillips he went to Harvard, both as an undergraduate and a graduate stu-
dent, and left Harvard in 1960 with a doctorate in classics and Sanskrit.
Thence, ultimately, to Brown in 1971.

Throughout his student and academic careers he racked up honors galore.
In 1981, he was awarded the prestigious MacArthur Fellowship.

David was a person who resided in a rather different mental and intellectual world from his departmental colleagues at Brown. To put the matter vulgarly, he was a horse of a different color. It was not just that as time progressed his major interests lay more and more to the east of Europe, Egypt, or Mesopotamia, but that in contrast to his departmental colleagues who cast the light of rationalism on their subjects, David saw the world as partly the end product of mystic, irrational forces. His perspective was located at the razor's edge of what is to be considered knowledge and what once was so considered, but now is not. His studies delved deeply into ancient ideas and actions that we would now consider laughable. Nonetheless, there was considerable overlap between his views and those of his more rationalistic colleagues, in particular on what scholarship consisted of.

I never had the guts to repeat to him Saul Lieberman's quip about Gershom Scholem's great work on the Kabala:* "Trash is trash; but the study of trash is scholarship." But let me conjecture what his reaction might have been. A soft-spoken and tolerant person, David would have laughed and then said "Yes, but there's always something of value to be learned." And indeed there is. David Pingree was a unique-un, or as they say in his profession, he was a *hapax legomenon*: one of a kind, and by his own admission an oddball who had a sharp nose for other oddballs.

When I first met David it was a case of instant friendship. Somewhat later, I used to banter with him that our friendship was due to the fact that we both grew up in Essex County, Massachusetts; he in Andover and I in Lawrence. I put out the claim that the air was much better in Essex County than in Middlesex or Suffolk counties. David nodded in agreement.

Once, speaking of Essex County (of which Salem is the county seat), he told me that he was the direct descendant of two Salem witches. I asked him whether he believed current sociological explanations for the Salem witch hunt, such as those you find in Arthur Miller's play *The Crucible*. He demurred, saying that his ancestors themselves believed they were witches and that was the end of the matter as far as he was concerned.

Hadassah and I used to see David, his wife Isabelle, and their daughter Amanda socially quite frequently: in our homes in Providence, and in their newly constructed summer home in Abington, Connecticut, where I can assure you we did not indulge in professional talk. There, in the country,

---

* Gershom Scholem. *Major Trends in Jewish Mysticism*. New York: Schocken Books, Inc., 1974.

the talk might center around the deer and the foxes that came to call or the neighbors who were tapping their maple trees for syrup.

As a sideline to his interpretation of ancient texts, David made a deep study of human folly. He reveled in it. He knew of hundreds of oddballs in the bygone world, men whose lives had a component of folly. He was able to pick out of drear, often censored and sanitized descriptions of lives—very much in the way John Aubrey or Lytton Strachey were able to—those flashes of folly or of eccentricity that add much to our reading pleasure. You and I—all of us—enjoy reading about such follies because you and I are, of course, quite sure of our own sanity and of the correctness of our ideas. Reading about these ancient loons makes us feel superior and comfortable in our own private worlds, in the secure knowledge that we have determined beyond refute what is the case and what is not the case.

David would speak to me of his oddballs almost as though they were his personal friends; his eyes would twinkle when he called them "weirdoes." But he also called them "very fine fellows," "very fine gentlemen," and "the salt and pepper of the earth." But what did he mean by the term "oddballs?" Mavericks? Independent thinkers? People with a touch of madness? Misguided folk? Cranks? Heretics? Schismatics? Deviationists? Flat-earth advocates? Fakers? Mythic characters? Folk whose strange or unusual behavior or thoughts raise one's eyebrows? People who, on the one hand had accomplished something we now consider praiseworthy, and on the other hand, had written or done something that we prefer not to discuss? Well, I shall not formulate a definition: perhaps after viewing some of the people residing in David's personal Oddballs' Hall of Fame you will be able to come up with your own definition.

What is folly? When seen against the common wisdom and practices of an age, follies tend to become softened. Beliefs now held firmly often began as follies. Or as Bertrand Russell put it: "Do not fear to be eccentric in opinion, for every opinion now accepted was once eccentric." And then again, who knows what beliefs and practices that even now we hold firm will appear as follies, stupidities, and blindnesses to our descendants?

I used to tell people—by way of a joke—that David knew everything worth knowing prior to 1670 and that his domain of knowledge extended from the Orkneys to Tibet. Later, getting to know him better, I removed these limitations on the date for he once gave me a long lecture on the theosophist Madame Helen Petrovna Blavatsky (1831–1891) and her "boyfriend" Col. Henry

Olcott. Blavatsky wrote *Isis Unveiled*, subtitled *A Master Key to the Mysteries of Ancient and Modern Science and Theology*. (It's still in print.) David added that Blavatsky sold ostrich feathers before she came to America. I mention these things as lagniappes in anticipation of the stories that are to follow.

What I will do in this short work is to take a number of the "weirdoes" or "oddballs" that David familiarized me with, go back to the reference works that he would always provide, and flesh out their stories just a bit beyond his teasing introductions. Digging into the primary or secondary sources (as opposed to encyclopedia or Internet write-ups) led me, an innocent, into worlds that I had not known: worlds of biography, history, and scholarship; worlds of iconography and paleography; worlds of primitive science and of strange, primitive beliefs and rites. One reference, one book, would often draw me into others until I had to say to myself "enough."

Of the 50 or so oddballs that David mentioned over the years in conversation or in letters to me, there was only one that I really knew something about: Pythagoras, and I suspect the same may be true of my readers.

In a number of cases, my introduction to these characters revealed to me the breadth of the individuals who have been (alas) epitomized by their foibles. These very foibles or follies were always my entering points and the *points d'appui* of my "professional" conversations with David, and a part of my sustained interest in the stories I now wish to tell. I feel sure that apart from their connection to David, revealing a little-known side of him, the stories will be read for their own intrinsic interest.

Yet, I found that digging out this kind of material was infectious; and I found myself chasing down to its origins some odd fact that I had run across myself. The chase would weave its way through my normal activities and concerns about applied mathematics and I intend here to do a bit of such weaving. I find it a pleasurable hobby and perhaps my readers will also be drawn to it.

Part II
The Loons and How I Got to Know Them

Simon Forman
Elizabethan Physician and Astrologer

I n the first months of our friendship, when we still were not on a first name basis, I received a letter from David Pingree through the university mail system.

> *Dear Prof. Davis:*
>
> *I understand you are a cryptographer. If so, I have an Elizabethan puzzle that might amuse you. Could you stop by my office some day?*
>
> *Sincerely,*
>
> *David Pingree*
> *Wilbour Hall*

Somewhere along the twisted path called "my career," I picked up the reputation of being a code breaker. This is not the case. I would be hard put to solve the Jumble in the morning paper. What is the case is this: I know a fair amount about the techniques involved. Since, in recent years, cryptography has become a severely mathematical and computer-involved subject, I have given courses on the topic, and some of my students have found jobs with the National Security Agency. Nonetheless, deserved or not, the aura of code breaker clung to me. The aura has long since departed.

Well, I stopped by David's office in Wilbour Hall—that dark red brick Victorian Ugly that I mentioned earlier. I knocked on the door of Prof. David Pingree's office. A man of about 40 opened it.

"Ah, I'm Pingree. We meet again. Splendid. Won't you sit down?"

I looked around David's room. Short of the library stacks, I had never seen so many books piled into a single room. Where could I sit down? Every square inch of horizontal surface was covered. Books, papers, notes, manuscripts—all congregated in random and chaotic disorder. The only available

David Pingree in his office

As the artist Marguerite Dorian imagined Prof. Pingree's office

chair was piled three feet high with the musty and moldering volumes of a decayed concordance. Underneath the chair, like Cerberus, a nondescript brown mutt eyed me with suspicion and guarded the concordance. David figured out my dilemma and brought in a chair from an adjacent classroom.

The previous summer Hadassah and I had been in London and seen *Brief Lives*, a brilliant one-man show performed by the famous actor Roy Dotrice. The play is a monologue delivered by John Aubrey, an eccentric 17th-century antiquary and accumulator of biographical oddities. The mise en scène was a masterpiece of gothic dust and chaos, books, minerals, specimens, armor, spider webs, and unwashed crockery.

"*Your office,*" I said to David, "*reminds me of the set for* Brief Lives. *Minus the spider webs, of course.*"

"*Oh, you like Aubrey?*"

"*Yes. Very much indeed.*"

"*Great man, John Aubrey. Great man. He cast spells to keep spiders away. I have spiders too, but the spells aren't effective.*"

"*I take it, then, you've tried out the spells?*"

"*Why not? Not much lost.*"

David went over to a bookcase and took down a book that had some pho-
tocopies stuck in the pages. He handed the sheets to me. What I saw were
lines of writing in a kind of script that was totally unfamiliar; a sequence
of dashes and odd scratches of the pen plus a figure that I recognized as a
horoscope. (One page of the document is reproduced below. The horoscope
is dated 1598.)

"I should like to get this deciphered," David said to me hopefully.

"This is in code?"

"I believe so."

"Who wrote it?"

"Simon Forman," David answered, and the tone of his voice indicated that,
of course, everybody in the world knew who Simon Forman was.

"Who is he?"

*"An Elizabethan physician and astrologer. These are probably sheets from a
diary or a case book of Forman's."*

A page of Forman's notebook (yet to be deciphered)

"Where'd you get it?"

"From the Bodleian Library. There's lots of Forman's stuff in the Bodleian. Some of it's been deciphered. There's a guy by the name of A. L. Rouse who thinks he can identify Shakespeare's Dark Lady on the basis of some of Forman's medical notes.[*] Forman had a practice among the aristocrats. Rouse wrote a book about it."

"Who was the Dark Lady?"

"Rouse thinks it's. . . . Well, read his book. It's lots of fun. But crazy."

"Why do you want to get this stuff deciphered?" I asked David.

"It may tell us a lot about medical and astrological practice 400 years ago.

Simon Forman

They were intertwined, you know. If you went to a physician 400 years ago, the first thing he did—that is, if you had money—was to cast your horoscope. Then he would take a urine specimen."

"Horoscopes today are not covered by medical plans."

"You're right. It may come to it, though. Simon Forman's cipher is used to conceal his sexual exploits in his annual accounting of unusual events in an attempt to rectify his nativity horoscope. Forman was a man who believed deeply in the astrological order of things. In the more extended passages in the BL, Sloane, 1822, they are connected with accounts of his casting talismans for magical purposes; in one case they are accompanied by a Latin prayer asking the angels to intercede on behalf of a virgin."

At this point in our conversation, I confessed and explained my deficiencies as a cryptographer. However, I said that I would send it to a professional cryptographer I knew.

"Abe Sinkov will probably make meatballs of this in ten minutes."

"I doubt meatballs. The writing, as you can see, is much faded. It's probably a transposition cipher, but it's full of short, telegraphic language, full of abbreviations, full of medical, alchemical, and astrological terms. And it's very likely written in a mixture of Elizabethan English, French, Latin, and Greek,

---

[*] A. L. Rouse. *Sex and Society in Shakespeare's Age: Simon Forman the Astrologer.* New York: Scribner, 1974.

Emilia Bassano Lanier

all of the pig variety. And one doesn't know where one language begins and the other leaves off. After all, it's a series of notes jotted down rapidly in a very personal cipher."

David proved correct as regards the difficulty of decoding. I sent the sheets off to Prof. Abe Sinkov, retired in Phoenix, Arizona. Sinkov kept the material for several months and then wrote me that he needed more material of the same sort to work with. In the meanwhile I read A. L Rouse's identification of Shakespeare's Dark Lady as Emilia Bassano Lanier. Take this identification or leave it.

As I write these lines, many years after the conversation just recorded, I believe that the residuary Forman material in the Bodleian remains undeciphered.

In Which I Meet Lord Dacre

The stories David told me led to personal consequences. This was inevitable. He familiarized me with the Elizabethan Simon Forman and with the contemporary English historian A. L. Rouse, and these introductions prevented me from being embarrassingly mute at a don's luncheon in Peterhouse College, Cambridge. I must begin my story somewhere, but where? How far back shall I go? Time seems to run linearly, but connections to people and to events get all mixed up in my mind, especially as time goes by.

Some years ago, I found myself in Cambridge, England, on a speaking mission. My college away from college, if one can call it that, was Pembroke. I was on friendly terms with Michael Powell, one of the Pembroke dons, and through him and the courtesy of the bursar, I arranged from time to time to stay for a few days in one of its guest rooms. On a few such occasions, Hadassah was with me. We would eat breakfast with the students and the resident faculty on long oak tables in the Hall, and the unchangeable menu that included stewed tomatoes, baked beans, eggs sunnyside, sausages, and fried toast remains paradigmatic in my mind of plain living and high thinking. This selection is often called the "full English breakfast."

Across Trumpington Street from Pembroke and a bit up is Peterhouse College, the oldest of the Cambridge colleges. I had never visited Peterhouse, despite the fact that the poet Thomas Gray, about whom I'd written a few lines, had once lived there. On this particular trip to Cambridge, though, I had a luncheon date with Jacques Hayman of the Cambridge University Engineering Department. Hayman had once been a member of my department at Brown and was now a fellow of Peterhouse.

Hayman showed me around the Peterhouse library and walked me through the ancient stones and narrow spiral passages that called to my mind an illustration in a beginner's history of England, in which the two unfortunate young princes, Edward and Richard, were dispatched while crouching in such a passageway. In my reverie, I updated the recollection by thinking that this was an excellent place for a murderous critic to slink up and dispatch one or two philosophers. Hayman pulled me out of the 15th century by telling me it was time for lunch in the dons' dining room.

"*How nice*," said Hayman, rubbing his hands together. He had looked around at the scattered assemblage of faculty in the dining room, and had spotted The Master.

"*I shall seat you next to The Master.*"

"*Great honor, I'm sure. Thanks. And who is The Master?*"

"*Lord Dacre.*"

"*Excellent. And who is Lord Daggers?*"

"*Dacre. D-a-c-r-e. You've probably heard his baptismal name: Hugh Trevor-Roper.*"

"*Of course. I've even read one of his books.*"

Hugh Trevor-Roper, Baron Dacre of Glanton, distinguished and controversial historian of Europe, who had moved lately from Oxford to take the mastership of Peterhouse, sat at the head of the long table, and I was seated at his left. After a few polite nothing-at-all words, during which I said I was a mathematician, he, the humanist, made the usual excuses that he was never very good at mathematics. I sought desperately in my brain for something in the line of history that we might talk about. I had recently read his *The Hermit of Peking*, a biography of the eccentric Sir Edmund Backhouse (and incidentally a very strong candidate for David's Oddballs' Hall of Fame). Among other eccentricities Backhouse had left behind a pornographic autobiography, which Trevor-Roper pronounced as pure imagination; but there at lunchtime, my mind went blank and I couldn't for the life of me think of the title or what it was about. All I could think of was that a few years before, in 1983, Trevor-Roper had authenticated a newly found diary of Adolf Hitler—which turned out to be fraudulent.

Hugh Trevor-Roper

Well, Hitler's diary would definitely not have been a friendly conversational gambit, so I scrounged around desperately in my limited historical knowledge for another topic. I came up with A. L. Rowse's book *Sex and Society in Shakespeare's England* in which, as David Pingree had told me, Rowse identified a certain Emilia Bassano Lanier as Shakespeare's mysterious and problematic Dark Lady. This conversational ploy worked splendidly. Rowse was a prolific writer on Elizabethan personalities and Trevor-Roper and Rowse were at daggers drawn professionally.

During our lunch (which was a substantial one), Trevor-Roper peppered our conversation with Rowse–gossip, to which I was able to respond and expand upon with some intelligence and not a little surprise. Rising from the table magisterially, he pronounced a final judgment that one must not take seriously the writings of an overheated brain. To which I nodded in weak agreement, for I knew that overheated brains led sometimes to great discoveries and sometimes to great enormities.

Hugh Trevor-Roper will appear later in this book in the sections on John Napier and Katharine Firth.

Elias Ashmole
Antiquary, Alchemist, Freemason, Historian, Founding Member of the Royal Society, and Benefactor

D avid was a great storyteller and had an unlimited supply of them. I soon latched on to this aspect of his knowledge to supply me with material that I might use for my fictional writings. His stories of eccentrics throughout the millennia were part of the light fringes of his vast supply of knowledge, and he shared them generously with me, revealing a playful side of his character.

Start him anywhere. He would then produce someone I and perhaps my readers had never heard of: say, Elias Ashmole. Well, I suppose you've all heard of the Ashmoleon Museum in Oxford. I had, and told David that that was the limit of my knowledge of the man.

David's eyes twinkled.

"Ashmole? A lively character. Very attractive, but a fruitcake. Great man, but a fruitcake. You know, he cured himself of ague by hanging three spiders around his neck. And then why did he work so hard writing about the Order of the Garter? Preferment? He was one of those astrologers who founded the Royal Society."

Elias Ashmole

"There may be something in the spider story for pharmacology to look into. You know," I said to David, reminiscing a bit, *"Neugebauer once told us at lunch a story that he'd picked up from an ancient Egyptian source. A man came to a priest/doctor with a complaint, and the doctor prescribed an application of crocodile droppings. . . ."*

"I remember the story," David interrupted me. *"And then The Elephant said to us: 'You see, the ancients knew all about hormones.' Neugebauer was having fun with us. He may have made up the story himself just to stress a political point."*

My recollections of what David told me about Ashmole (1617–1692) are now a bit dim, but one thing has stuck. In the 17th century, there was hardly any distinction between mathematics, physics, alchemy, astronomy, astrology, numerology, magic, theurgy, or Kabala. All these disciplines were rolled together in one unified interpretation and practice, and its practitioners were often regarded with more than a little suspicion that often verged on the heretical. The Royal Society, now one of the leading scientific societies in the world, was founded by astrologers (and I have been one of the editors of its *Transactions*).

Somewhat later, I came across a nice quote from John Maynard Keynes that bears on this paradox:

> [Sir Isaac] Newton was not the first of the age of reason. He was the last of the magicians, the last of the Babylonians and Sumerians, the last great mind that looked out on the visible and intellectual world with the same eyes as those who began to build our intellectual inheritance rather less than 10,000 years ago.

I find in my notes a memo that David sent me, consisting of a clip from what appears to be Ashmole's diary. His sending it to me may have been in response to my questioning him on the use of sigils. (A sigil is a sign or a word or some kind of device that is held to have occult or magical power; cf. amulet, talisman. Books of sigils are still being published today and amulets abound.)

17th cent. Elias Ashmole
Ashmole 431 ff. 103-188

> *The horoscopes of the sigil-casting say nothing of the form of sigils except f.137. The Figure of a Ratt was cast in Lead and made in full proportion, but had noe Character upon it.*
>
> *The Figure of a Mole was cutt in an Ovall Figure upon a Punchion of Iron, lying longe waies over the back were these Characters set* [two strange figures here] *under the belly these* [two strange figures here].
>
> *And they were stamped upon Lead.*
>
> *The Figure of the Caterpillare & Flyes Fleas were all made in full proportion, in litel, & cast off in Lead without Characters.*

I once remarked to David at lunch that there was a definite anti-rationalist spirit in the air today: that irrationalism was making serious inroads; that Kabala was back; that Mrs. Reagan, the wife of the president, consulted her

An assortment of sigils

astrologer regularly and that was the least of it. I suggested to David with a twinkle in my eye that, with his knowledge of all the ancient arcane irrationalities, he could make a lot of money by becoming a magus. David put off my suggestion with a grunt:

"Irrationality has always been around and always will be. Just look around and you'll find sigils all over the place. Good luck symbols. In automobiles, horseshoes over doors, on walls, talismans around necks, eyes painted on the prows of oriental boats, four leaf clovers. The number of books in print that promote irrationalities is enormous. Not so many centuries ago, some of what is promulgated would have led the authors and booksellers directly to the stake."

On that note, we parted—he to go back to Wilbour Hall, and I to my office to prepare my next day's lecture.

Thomas Allen
Mathematician, Astrologer, and
Reputed Practitioner of Black Magic

As I learned from David and as I have already mentioned, in the Elizabethan age magic and spells were widely practiced and mathematicians often had more than a finger in this practice. Thomas Allen (1542–1632) is reputed to have been the best Oxford mathematician in the second half of the 16th century. That may be the case. He was an indefatigable collector of old manuscripts, but by today's standards he has left behind little by way of new theorematic mathematical material.

I think that David's interest in Thomas Allen may have been in the fact that Allen was connected tangentially with one of the

Thomas Allen

Amy Robsart

most famous unsolved mysteries in British royal history: the presumptive murder of Amy Robsart, Lady Dudley. (The death of Princess Diana has been "settled" via the Lord Stevens' report.) The demise of Amy Robsart (1532–1560) has been the subject of conjectures, paintings, novels (e.g., Sir Walter Scott's *Kenilworth*), plays (by Victor Hugo and by Eugene Scribe), and an opera (Daniel Auber)—and, with the exception of *Kenilworth,* none of these has been very successful. Amy also appeared recently in a short segment of the *Masterpiece Theater* TV production called *The Virgin Queen*. She was played by the actress Emilia Fox.

Here is the story very briefly: Robert Dudley, the Earl of Leicester, married Amy Robsart, much against the desire of Queen Elizabeth I, his sovereign (who, some have said, had the hots for him). Dudley was fond of Thomas Allen, and Allen, who had the reputation of being a magician, worked black magic to assist whatever hanky-panky Dudley was up to and also to foster a match between Dudley and Elizabeth. So it was rumored, anyway.

One day Amy was found at the bottom of a staircase with her neck broken. Rumor had it that Dudley hired a hit man to dispose of Amy so he could be free to marry Elizabeth.

And I'll let my readers take the story from there. Read *Kenilworth* or watch any of the productions in which Amy appears.

Siva and Parvati
or
How the Sun Married the Moon

O ne year I spent several academic quarters in Salt Lake City teaching at the University of Utah. I wrote to David on January 7, 1980, as follows:

I have just stumbled across the ascetical works of Gregory of Nyssa. He gives all the wrong reasons for joining a monastery and remaining celibate. I need a quick antidote. Prescribe.

Yours, PJD

David's response came rapidly.

January 16, 1980

Dear Phil: You are right. For such as us, Gregory may be a bit hard to take. You can redress the balance by reading how the Indians connect yogic practices with eroticism.

Read The Birth of the War God, *a poem by Kalidasa (c. AD 400). This tells how Siva practiced the asceticism way of protest and destroyed herself. She was then*

49

reborn as the Daughter of the Mountain. This rebirth was vital because the creative powers of the universe depended on her. It was necessary for her to overcome the asceticism of the god Siva and this she did by practicing asceticism of her own.

Sincerely, David

I was confused by the sexes of the individuals mentioned in David's letter. (Later I found out that Siva is androgynous.) Poking around in the Marriott Library of the university, which I found remarkably good, I turned up a slightly more elaborate version from the *Kumarasambhava* of Halides entitled *Siva and Parvati*. I went through it as I used to go through *Grimm's Fairy Tales* and in my mind I gave it the subtitle *How the Sun Married the Moon*. Here is the way I figured my way through what was a very complicated situation.

Siva was married to Sati, the daughter of Daksa. There was an insult exchanged between Siva and his father-in-law Daksa. Sati, in despair, practiced asceticism, and she was so successful at it that her body wasted away. But not to worry. Her soul (atman) was reborn as Parvati, the Daughter of the Mountain (Himalayas). In the meantime, Siva, sorrowing over the loss of Sati, abandoned the world, practiced yoga, and entered into deep contemplation (samadhi).

Now there was a certain very evil fellow by the name of Taraka. Taraka had been practicing yoga and had thereby gained enough "heat" (tapas) that

Siva and Parvati

he was able to blackmail Brahma. Taraka threatened to burn the universe, and the price he asked was to be installed in heaven in control of the gods and the rest of creation. The gods were told by Brahma that they could get rid of Taraka only by finding a young leader, Kumara (who would be the son of Siva and Parvati). The gods sent the god of love, Kama, to cause Siva to fall in love with Parvati. Siva (the Sun) burned up Kama with the fire from his third eye. But Parvati (the Moon) had, on her own, decided to marry Siva and she practiced asceticism to gain her end.

Where, as Sati, her asceticism had destroyed her body, as Parvati, it awak-
ened Siva from his asceticism. They were duly married, Kumara was born,
and the universe was saved. Now go figure out this one.

Pythagoras
Mathematician, Philosopher, Mystic, and Organizer of an Ascetic Religious Brotherhood

I needed more reading material. I wrote to David that his recommendation of Khalidasa was an excellent choice but that my bedtime reading was still suffering from a lack of lurid material. The *Deseret News*, the local paper, did not engage me. David immediately wrote back a long letter, which I will abridge here but draw on later.

> *6 March, 1980*
>
> *Dear Phil:*
>
> *At last I drag myself from beneath the ever-growing pile of half-eaten and thoroughly undigested processed wood pulp (certainly less comestible than papyrus, parchment, or rag), and cast pen to an untouched sample.*
>
> *I am delighted that the Latter-day Saints are interested in their predecessors, at least to having on their shelves that storehouse of human folly, the* Acta Sanctorum. *But really, for high amusement, you should try to get hold of a translation of Aelius Lampridius' Life of Heliogabalus.*
>
> *I assume you have read as well, Philostratus' Life and Apollonius of Tyana's own* Letters. *Read Iamblichus on Pythagoras. Besides these you must read the account of*

Apollonius' magical acts in Greek and Arabic given by P. Krauss in volume 2 of his Jābir ibn Hayyan. *Jābir incidentally, should join our list of weirdoes.*

And so on for several more handwritten pages. I'm sure that my interest in the curiosities of the deep past stirred up his knowledge of them and telling them to me constituted a relaxation from his professional studies and his writings.

I didn't get to Pythagoras until I was back from Salt Lake City and in Providence. We had a conversation about these "fine gentlemen." David had recommended that I start with Iamblichus' *Life of Pythagoras* (*De Vita Pythagorica*) and I asked him whether Iamblicus knew Pythagoras (c. 582 BCE–507 BCE) personally. (What a naive question this was!) Not at all, was David's answer, Iamblichus was a third century (CE not BCE) neo-Platonist writing eight centuries after his subject. I knew a bit about neo-Platonism. I had had some conversations with the brilliant Chassidic scholar Abraham Heschel, who suggested that I read Plotinus. I read a few pages but Plotinus wasn't my cup of tea.

PD: *Eight hundred years? That would be like my sitting down and writing a biography of Dante!*

DP: *Right. And this kind of thing is done every day of the year.*

PD: *I wonder how Ronald Reagan will come across 800 years from now.*

DP: *Hah!*

PD: *So where did Iamblichus get his info?*

DP: *There was lots of material around. Diogenes Laertius. Apollonius of Tyana. Other guys. They all wrote bits and pieces about Pythagoras. Now speaking of Apollonius, there was a fruitcake if ever there was. You should get onto him.*

Iamblichus

PD: *Will do. But where's Tyana?*

DP: *In Cappadocia.*

PD: *Ah, so.* [I pretended I knew where Cappadocia was.] *And what's in Iamblichus?*

DP: *Nothing, really, about math in his vitae. A bit about his life, and mostly about the rules he set up for his society. The way to live properly and decently. Take care of both your body and your mind. While you're at it, take a look at his* De Communi Mathematica Scientia (On the General Principles of Mathematics). *More stuff there.*

PD: *Will do. I remember that Bertrand Russell wrote that Pythagoras founded his society on the basis of the transmigration of souls and forbidding his followers to eat beans. Beans!*

DP: *That's Russell all over. Most of the rules of the Pythagorean brotherhood were about temperance, decent social behavior. If we could only follow them, it would lead us to a better world today.*

PD: *As regards the beans, I suppose Russell was pretty stupid. Beans lead to flatulence, and, as you've said, Pythagoras was concerned with purity.*

Bertrand Russell

DP: *Rationalist explanations are only part of the story. How can they account for taboos about beans that have nothing to do with eating them?*

PD: *Such as?*

DP: *Such as not walking in bean fields when the beans were in blossom? Or that the souls of people were in beans?*

Every high school student of geometry knows—or should know—the theorem of Pythagoras. It says that the area of a square built on the long side of a right-angled triangle equals the sum of the areas of the squares built on the two shorter sides. In algebraic symbols that were not around in Pythagoras' day: $C^2 = A^2 + B^2$.

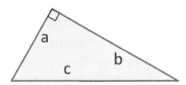

Somewhat more sophisticated students know that Pythagoras is reputed to have proved that the square root of two ($\sqrt{2}$, and some also say $\sqrt{5}$) is irrational. This means that an equation of the sort $\sqrt{2} = a/b$, where a and b are whole numbers, can never be true. This is supposed to have thrown Pythagoras into a geometrical/logical tizzy, for $\sqrt{2}$ is the length of the long side of a right-angled triangle whose two shorter sides both have length 1. Thus $\sqrt{2}$ exists and yet does not exist because whole numbers and their ratios

Pythagoras

were the only numbers available on the ancient Greek mathematical palette. This dilemma—a later worry—is called the Crisis of Pythagoras and is the first of numerous crises in the history of mathematics.

Legend has it—and Pythagoreanism is nothing if not full of legends, often contradictory ones—that Pythagoras was so amazed and bewildered by his discovery that he sacrificed a hecatomb of oxen to the gods and his followers were sworn to silence about the matter. Listen to the reaction of mathematician Charles Dodgson (Lewis Carroll) to this story from *A New Theory of Parallels*:

> Pythagoras celebrated the event, it is said, by sacrificing a hecatomb of oxen—a method of doing honor to science that has always seemed to me slightly exaggerated and uncalled for. One can imagine oneself, even in these degenerate days, marking the epoch of some brilliant discovery by inviting a convivial friend or two to joining one in a beefsteak and a bottle of wine. But a hecatomb of oxen? It would produce quite an inconvenient supply of beef.

What other pieces of mathematical Pythagorianisms does the history of the subject tell us? Some have conjectured that he probably knew the geometrical contents later enshrined in the first two books of Euclid's *Elements* (third century BCE). That he knew various properties of the regular pentagon. The pentagram was the logo of the Pythagorean brotherhood and it was believed that the very figure contained mysteries wrapped up in further mysteries.

The basic pentagram

Pythagoras (or the Pythagoreans) divided numbers into the masculine and the feminine. They attributed various magical/mystical properties to several of the whole numbers, particularly those connected with geometry: the triangle, the square, arrangements of pebbles, etc. Pythagoras discovered and philosophized about the harmonic ratios of the lengths of vibrating strings. Thus 1:2 is a diapason, 2:3 is a fifth, 3:4 is a fourth, etc.

There is geometry in the humming of the strings, there is music in the spacing of the spheres.

Here we have the very first instance of the "music [or the harmony] of the spheres." In fact, it was Pythagoras who coined the words philosophy and harmonic.

Pythagoras said "All is number." But was mathematics to Pythagoras simply the tip of his intellectual pyramid or was his

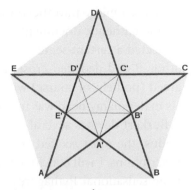

A pentagram within a pentagram

saying "All is number" an indication that numbers, through numerology and number symbolism, was the thread that bound together his piety, his philosophy, and his rules for a life properly lived? The question of how to bridge the gap is still discussed today.

I first learned about Pythagoras through mathematics—in high school or slightly before. Then there was a hiatus of about ten years until November 16, 1945, when I bought a copy of Bertrand Russell's newly appeared *A History of Western Philosophy* and inscribed the date therein. I will remind readers that Russell (1872–1970), throughout his long life, was tremendously prominent as a mathematical logician, a philosopher, a social moralist, and a political agitator. He was a beautiful writer, witty and prejudiced, and his history is not 100% reliable. Russell opened his chapter on Pythagoras with the sentence

> Pythagoras . . . was intellectually one of the most important men that ever lived, both when he was wise and when he was unwise.

(Russell, in parallel, was one of the most prominent intellectuals of the 20th century, both when he was wise and when he was unwise—which was not infrequently.) A bit of an oddball, certainly, and if David had not cut off the lists he gave me at the early 19th century, he surely would have included Russell.

Yes, Iamblicus in his *Life of Pythagoras* only hints at all these things. Even the famous theorem is glitched there—what comes close to it is a brief mention of the 3-4-5 right triangle in a political context! Tantalized, I went on to other books on Pythagoreanism and was led rapidly to the magisterial study of Walter Burkert. This 500-page study made me realize that Pythagoreanism is not merely an ancient doctrine, it's an industry that has engaged scholars since Pythagoras shook off his mortal coil. Burkerts' work is a book that threw me rapidly into the world of ancient authors and authorities, spurious and authentic documents, a plethora of long footnotes in which scholars connected a certain sentence of ancient author A with a certain sentence of author B and deduced cross-fertilizations, and where the interpreters disagreed (politely, of course) with each other. I could not handle or absorb all this material: my head was in a whirl.

Burkert devotes several pages to the issue of the beans, surely the most famous, the most scoffed at, and the most mysterious of the Pythagorean proscriptions. I will conclude this report of my reading by listing a few more of the sensational Pythagorean *acusmata*, or orally transmitted maxims or rules of operation.

- The most just thing is to sacrifice to the gods; the wisest thing is number; the most beautiful is harmony; the strongest is insight; the most desirable is happiness; the most beautiful shapes are the circle and the sphere; the truest thing is that men are wicked; the holiest thing is the mallow leaf.

- One should not sacrifice a white cock.

- A man should not have children by a wife who wears gold jewelry.

- Do not eat bread from a whole loaf.

- Do not stir the fire with a knife.

- Do not step over a yoke.

- Do not eat the fish that are sacred to the gods.

Priscillian

Magician, Astrologer, Heretic, and the Only Christian Heretic to Be Executed by the Civil Authorities in Antiquity

A s I've said, during the time I spent in Salt Lake City, the world center of the Latter-day Saints (LDS), David sent me long lists of characters whose stories, he asserted, would keep me well supplied with bedtime reading. He divided his characters into four categories: politicians, scholars, religious fanatics, and saints—the last two categories overlapping. His list of saints included such men as Cyril of Alexandria, Priscillianus, St. Cuthbert, and St. Christopher the Dog-Faced.

David ended his list with these words:

> *Let me know what you think of these fellows and what other sorts ought to be thrown in. The list could be expanded almost to infinity; the only limiting factor is the relatively short time that men have suffered the curse of literacy.*
>
> [His writing became faint.] *As you see, my pen is becoming illiterate with extreme rapidity. Therefore I desist.*
>
> *Sincerely, David*

In all the years of our friendship, we never spoke of our personal religious beliefs or practices, but I would infer that he would agree with the following assessment—and here I parallel the famous sentence of Edward Gibbon—that to liberal governments all religions are equally true, to philosophers they are equally false, and to scholars such as himself they are equally useful as grist for their descriptive and interpretive mills.

Let me take up the case of Priscillian (d. 385), far from a saint, but in his day thought by some to be fairly close to one. Today Priscillian is hardly on the tongue-tips of the devout. He is grist only for the mills of scholars of early church history who occasionally mull over the copious supply of still-existing and relevant documents. But David remembered him and inscribed him among his ancient oddballs.

I've been to a number of places where a plaque on the wall advised on-lookers that on this spot such and such a person lost his life. I've been to the spot in Canterbury Cathedral where in 1170 St. Thomas à Becket was hacked to death by four knights, thus ridding King Henry of a meddlesome priest. I've seen the spot in the Holyrood Palace in Edinburgh, Scotland, where in 1566 David Rizzio, the private secretary and a presumptive lover of Queen Mary of Scots, was stabbed to death.

I've even seen the sign on Route 375 in Israel, in the Valley of Elah, that says here it was that the young David dispatched the giant Goliath with stones from his slingshot. But in Trier, Germany, I did not see a sign that said here, on this spot, Priscillian, Bishop of Avila, was judicially murdered in 385. The fact was not even mentioned in the tourist pamphlets.

The city of Trier (Treves) is located on the Moselle in the wine-growing area of the Rhineland-Palatinate. Trier is an old city dating back to Roman times, and possesses considerable Roman residue to look at and to ponder. Some years back, I was a speaker at a conference in a village about 20 miles from Trier, where the German Computer Society had converted a baroque palace into a conference center. One afternoon, during a break in the conference, I joined several of the conferees and drove with them to Trier for a bit of sight-seeing. I recall a large well-preserved amphitheater easily seating 20,000. We were allowed to go into the bowels of the amphitheater and visit what might today be called the greenroom of the gladiators and criminals. The lions' cells were adjacent, of course. Notable also was a huge, heavy, ugly structure called the Porta Nigra (the black gate), which over the millennia has been used as a Roman administrative office, a church, and today as a ticket office for tourists.

The Aula Palatina in Trier

The Aula Palatina, a basilica dating from around 310 (now restored and serving as a Lutheran church), contained the assembly hall and throne room for the emperors and high government officials. A tourist attraction of a more recent vintage, pulling in both the devout and the curious, is the house in which Karl Marx was born, now set up as a museum and displaying a modest amount of Marxiana. Marx did not interest me at that moment, but I went along for the sake of group solidarity.

Now I return to the fourth century and to Priscillian, to flesh out David's brief introduction. The barbarian tribes were pressing the Roman world on all sides. The Roman Empire, split into the East and the West, was in a state of disintegration. Christianity was in the ascendant and, according to some historians, had more than a finger in the dissolution of the Empire. The emperors squabbled with one another—often with lethal power grabs. The bishops and theologians quarreled with one another over both turf and doctrine. They anathemized and hopefully extirpated heretical beliefs by the dozens and cobbled together a more-or-less unified dogma that followed the lines laid down by the Council of Nicea (325). It is said that the Emperor Constantine, who suggested the Council, would have been pleased with whatever the theologians came up with as long as agreement was achieved. Sounds like a piece of wisdom that's in short supply today. You can believe that the

Emperor Magnus Maximus

history of the early Church is a very well-ploughed field, with a myriad of ancient documents and opinions cited by historians—leaving me, a complete outsider, thrashing about in a sea of footnotes.

Magnus Maximus (d. 388), emperor of the West for a few years, was a general who had been in charge of military operations in Britain (against the Picts and the Scots). He had his governmental seat in Trier, very likely in the Aula Palatina. Magnus was a devout Catholic and the scourge of pagans, unbelievers, and heretics (i.e., anyone who didn't hew to his line of belief). Were shows still taking place in the amphitheater with such misfits as victims? Were deviant Christians food for the lions? I don't know. The Trier brochure for tourists informs us that today, rock concerts and suchlike take place there.

As Magnus Maximus sat on his throne, a man came before him with a remarkable plea that today we would call a plea for the separation of church and state. His name was Bishop Martin of Tours, now St. Martin (d. 385), and he was received privately by Maximus. A certain Priscillian, a Spaniard and bishop of Avila in Spain, had the wrong—condemnable—ideas, theologically speaking, and a class-action suit against Priscillian, charging heresy and other peccadilloes, had been brought before the emperor. The class that brought the action was a group of concerned bishops from Iberia. Martin heard about this and went lickety-split from Tours to Trier to argue that, though he was dead set against what Priscillian and his followers believed, the case was an ecclesiastical matter and should not have been brought before the civil authorities. He cited as a proof text the opinions of St. Paul on the topic (2 Corinthians 6):

> Do you not know that your body is a temple of the Holy Spirit, who is in you, whom you have received from God? You are not your own.

At first, the emperor agreed with Martin. Martin packed up and went home. Then a certain Bishop Ithacius, the head of the group that was out for blood, was furious and went back to the emperor. Magnus Maximus, agreeing apparently with the last person with whom he spoke and having political reasons for not appearing soft on heretics, changed his mind and ordered

Priscillian and a few of his followers to be beheaded. This was in 385. Three years later Magnus Maximus, considered an upstart by Theodosius, emperor of the East, was defeated in battle and was himself executed.

What, precisely, were the charges brought against Priscillian, either in the trial or in the vast quantities of gossip that surrounded him? He was charged with being a magician, a sorcerer, a Manichaeist (i.e., a believer in a dualistic theology); it was said that he studied heretical, extra-canonical texts, that he indulged in nocturnal magical and sexual orgies, and that he was a Docetist (i.e., one who asserted that Christ only *seemed* to suffer). It was said, furthermore, that he advocated a Patripassian doctrine of God (i.e., he denied the independent preexistent personality of Christ) and that he was a bit of a feminist, believing that both men and women were equal in their spiritual faculties. For this reason Priscillian attracted a substantial following among women.

Some assertions are easier to demonstrate than others. In recent years, a person may be brought up on a dozen or more charges of corruption and convicted of one or two. Al Capone was finally brought to justice on an income tax rap. So it was with Priscillian. In his day, magical procedures were rife—and I don't mean pulling rabbits out of hats. The maleficent potency of magic was believed in and feared. Laws were passed against it and the practice of black magic became a capital offence. Priscillian was said to have participated in pagan fertility rites in fields of Andalusia—a no-no to the orthodox. After a substantial and complicated trial—with all the theological quibbling brushed aside—Priscillian was convicted of sorcery, condemned by the emperor, and executed together with a number of his followers.

Following the death of Priscillian, a hue and cry went up for a variety of reasons. For one thing, the trial was illegal (as Martin of Tours had demonstrated). But the principal source of unrest was the fact that Priscillian was a charismatic figure with a strong magnetic appeal to thousands. When news came of the deposition and death of Emperor Maximus, the Priscillianists were triumphant. Priscillian was declared a martyr by them and his grave became a shrine for pilgrims; he had followers well into the sixth century.

St. Christopher the Dog-Faced

O ne day at lunch, David came out with

"*I assume you know the word 'cynocephalic.' Right?*"

"*Wrong. No. I know some strange words such as 'Muggletonian.' I don't know cynocephalic. So tell me.*"

"*Very easy. Two simple common Greek roots so you could figure it out for yourself. But it means having the head of a dog.*"

"*I suppose all dogs are cynocephalic. Ha ha.*"

"*More than that, St. Christopher had the head of a dog.*"

My eyebrows shot up.

"*Did he now? Saint Christopher? Saint Christopher of the Saint Christopher medals? The patron saint of travelers? The saint who was de-sainted* [decanonized would be the proper term] *in the 1960s?*"

"*The same guy,*" David answered me. "*But Christopher has a split personality, the first deriving from the Roman Catholic tradition and the second from the Eastern Orthodox tradition. The second is the more interesting. And while*"

Anubis

Bast

you're at it, check out the idea that St. Christopher derives from the Egyptian god Anubis, the jackal-headed—an amusing transformation."

"I suppose there is also a term 'gatocephalic,' meaning having the head of a cat—if I have the Greek root correct. And how about," I added, proudly displaying my deep knowledge, *"there's the Egyptian goddess Bast who was often depicted as a woman with the head of a cat."*

"Right. Anyway, check out St. Christopher, the dog-faced."

What I found when I had time to look into the matter (it took me a few years to get around to it) is that St. Christopher legends abound. More generically, in the sense of David's abiding interest in the interchange of cultures, I found a book on the topic by David Gordon White, *The Myths of the Dog-Man*, a spin-off from White's doctoral thesis at the University of Chicago. What I found in White was a folkloristic mélange of dog-stars, dog-days, hordes of barbaric dog-headed men attested to by the ancient geographers, Christian legends, the role that dogs played in old Chinese burials, and much, much more. Even Prester John gets into the cross-cultural act.

I will not end my encounter with the dog-faced Christopher without including a bit of the legend material. The stories of our hero reached Ireland and this produced the Irish *Passion of St. Christopher*. I quote from the rendition of David Woods, a classicist of University College, Cork, who based his treatment on a previous work by the renowned folklorist Sir James Fraser of *The Golden Bough* fame.

> A certain Baceus went to the king [a pagan, as you will see], and said: "Hail O King, I have news for you. I have seen a man with a dog's head on him, and long hair, and eyes glittering like the morning star in his head, and his teeth were like the tusks of a wild boar. I struck him for he was cursing

the gods; but he did not strike me, and said it was for the sake of God that he refrained. I am telling you this in order to know what is to be done with him, for it seems that it is by the God of the Christians that he has been sent, to help the Christians."

"Bring him to me," said the king.

The bystanders said that a large number of men must be sent for him. "Let two hundred soldiers go for him," said the king, "and bring him hither in chains; and if he resist you, bring his head with you that I may see it."

The soldiers then went to seek him. As for Christopher he went into the temple, and drew his hair round his head in two plaits. He rested his head on his knee, and, after planting his staff in the ground, began to

St. Christopher the cynocephalic

pray. "Almighty Lord," he said, "perform a miracle through me that thy name may be praised; and let this staff send forth shoots." The staff immediately put forth twigs, and leaves and flowers appeared.

I leave this encounter at this pleasant high point because Christopher's fortunes went downhill afterwards.

Every once in a while I think of the word I coined: "gatocephalic." The smash Broadway musical *Cats* was still some years down the pike. If the theater reviewers had known this word, they could have talked about its talented gatocephalic singers and dancers and displayed their literacy in Greek.

St. Cuthbert

I recall the following conversation with David in his office in Wilbur Hall:

DP: *Read about Cuthbert and the onions.*

PD: *Cuthbert? Who is he? Oh yes, I think you recommended him sometime ago.*

DP: *A monk. A very important guy. Lived on an island off the east coast of northern England. Lindisfarne. It's a holy island, you know. Monastic living and all that.*

PD: *I didn't know. But did this guy hate onions? I, personally, can't imagine life without onions.*

DP: *On the contrary. You'll see.*

PD: *Aah. Cuthbert knew his onions.*

DP: *He did indeed.*

Time passed, but later, in his letter of March 6, 1980, David sent me an elucidation and a further enticement:

> *Cuthbert is the early miracle-monger whose life was recorded by Bede, Reginal, Symeon, and many, many others as well as by Anonymous. I chose him as a representative*

of the wild Northumbrian-Irish breed of ecclesiastics (for whom see Anatole France's Penguin Island*) because of his encounters with demons (especially those possessing ladies; cf. William of Auxerre [1150–1231], theologian who introduced Aristotle into Catholic doctrine) and his miracles. Incidentally, a good introduction to saintly miracles is C. G. Loomis,* White Magic, *Cambridge, Mass. 1948. I like Cuthbert's survival during his last illness on the nourishment derived from sucking on an onion.*

Incidentally, *Penguin Island* was one of Otto Neugebauer's favorite books.

In Cuthbert's day, Lindisfarne was isolated and remote from Europe, particularly from Rome. Nonetheless—and it surprises me—communication was possible. Perhaps the monks of Lindisfarne even received mail twice a day as we used to when I was a child.

Today Lindisfarne, the Holy Island, attracts both the average tourist and the devout. You will find there reception centers (sorry, centres), churches, retreats, hotels, B and Bs, and gift shops selling dishes, facsimiles, the works. Overall, an aura of piety hovers.

A few years back, Hadassah, Teddy, and I took a trip to Cornwall, making our base in Padstow. Once there, we had, of course, to drive to nearby Tintagel where a fortress is reputed to have housed the legendary King Arthur and the Knights of the Round Table. "Downtown" Tintagel was full of tourists—busloads of them from foreign countries—and the souvenir shops sold plastic swords and shields for the kids. It's just possible that the gift shops in Lindisfarne sell plastic miters and croziers, but I suppose that would be inconsistent with the air of devotion that pervades the island.

Now fast backward to the eighth century. Let me propose a fantasy: imagine that Cuthbert (634–687) sent to the Admissions Office of the College of Saints a CV that stressed the following:

- I became lame with a swelling of my knee and an angel cured me.

- I changed the direction of the winds by prayer.

- I cast out the devil that was residing in the prefect's wife.

- I talked the birds into staying away from a new sown field. Crows came and apologized to me.

- I saw the soul of a man who had been killed in a fall ascend to heaven.

Cuthbert could have produced many more such accomplishments, but these five, I conjecture, would have sufficed to gain Cuthbert early admission. Being a modest man, he didn't want overkill.

Of course, this is not the way things are done. One simply doesn't make a plea for one's own saintly status. People gasped when Napoleon crowned himself emperor in the presence of the pope.

Statue of St. Cuthbert

Think of an actor awarding himself an Oscar. (Of course, I've heard of authors setting up claques to push for Nobel Prizes for themselves.)

I don't know how things were in the eighth century, but today beatification and canonization are very complicated matters. The position of *Advocatus Diaboli* (the Devil's Advocate, i.e., the man who is supposed to challenge the claims of miracles, etc.) was not created until the late 16th century. And in the recent past a number of well-established saints have been decommissioned—or decanonized to express the matter more accurately.

It was not easy to find the list of Cuthbert's miracles. Nothing so specific is mentioned in the write-up in the old *Encyclopedia Britannica*, but a bit of scrounging produced it. On the other hand, it was very easy to get into the story of Cuthbert and the onions. Here is the story in the very words of the Venerable Bede (673–735).

> Seeing that he kept silence, I said, "I see, my lord bishop, that you have suffered much from your complaint since we left you, and I marvel that you were so unwilling for us, when we departed, to send you some of our number to wait upon you." He replied, "It was done by the providence and the will of God, that I might be left without any society or aid of man, and suffer somewhat of affliction. For when you were gone, my languor began to increase, so that I left my cell and came hither to meet any one who might be on his way to see me, that he might not have the trouble of going further. Now, from the moment of my coming until the present time, during a space of five days and five nights, I have sat here without moving." "And how have you supported life, my lord bishop?" asked I; "have you remained so long without taking food?" Upon which, turning

up the couch on which he was sitting, he showed me five onions concealed therein, saying, "This has been my food for five days; for, whenever my mouth became dry and parched with thirst, I cooled and refreshed myself by tasting these." Now one of the onions appeared to have been a little gnawed, but certainly not more than half of it was eaten.

You know, as David pointed out to me, the desire to provide rational explanations for everything in the world that seems odd is very strong. But it rarely washes. And so in a number of places I read that in the hot countries, raw onions are still used for the purpose of relieving thirst. An interesting explanation, but I hadn't heard that sliced onions are offered on hot summer days in Fenway Park to allay the thirst of Red Sox fans.

The desire for rational explanations contradicts the idea, current in contemporary scientific thought, that random, unexplainable events exist. But I'll leave a discussion of randomness for a later occasion.

In 793, the Danish Vikings raided the monastery in Lindisfarne, ushering in 400 years of Danish plunder, conquest, dominance, colonization, cultural and linguistic syncretization, etc. The body of St. Cuthbert, by this time declared a holy relic, was spirited away. It was brought here, it was brought there, and finally, after years, came to rest in Durham Cathedral.

Ten or so years ago Hadassah, her cousin Teddy, and I—on another trip: a tour in Yorkshire—visited Durham Cathedral and had lunch in its coffee shop. I was unaware that St. Cuthbert was so close by. So close and yet so far: for some say he is resting behind the high altar, while others say that only a few of the cathedral priests really know the exact location and pass on the information to their successors in their last moments. Well, whatever.

Arthur and Excalibur

When we returned to Providence, I told David that we had found ourselves close to Cuthbert.

"Splendid. Splendid. Does the gift shop sell souvenir onions?"

Our trip to Durham reminded me of our previous trip to Tintagel.

"Ah. King Arthur. Malory's story. You know, Sir Thomas Malory was quite a guy."

"I didn't know."

"In and out of jails. For burglary; he stole sheep. Robbed churches. Rape. So they said. He wrote Le Morte d'Arthur *in jail. Definitely belongs among the loons."*

I didn't get to the full Malory story until many years later, and I haven't included it in this book.

Virgilius Maro Grammaticus
(Vergil the Grammarian)

I n Salt Lake City I was getting bored—my teaching kept me busy during the day but what to do at night? Hadassah was back in Providence, my Utah mathematical friends were with their families, etc. I wrote to David and he replied immediately. It was an emergency. I have lost track of the exact date.

Dear Phil: You are not bored. You are in Heaven and you're unaware of it. In Heaven nothing ever happens. Ever. Many authorities attest to this.

But if this is too severe a judgment on Salt Lake City, let us say you are in Heaven's antechamber.

Make use of local resources. Look into the Mormons, my Dear Sir. Look into Joseph Smith. Look into Brigham Young. Go back to the Mormon Patriarchs, for I suspect that contemporary Mormonism has been sandpapered down a bit. When you have exhausted Mormon sources, or they you, jump back to antiquity.

The Merovingian Court provides a splendid example of depravity. Gregory of Tours in his History of the Franks *does not do justice to Fredegund. But there are other Merovingian sources (all in the* Monumenta Germaniae Historica*) which fill out the picture somewhat.*

A recent edition of a somewhat mysterious author

Do you know about Virgilius Maro Grammaticus who lived in Merovingian Gaul? He fought over the burning issue of whether, in Latin, the word 'ego' has a vocative. Recall: the vocative is the grammatical form in which someone is addressed: Oh, Caesar! Oh Cleopatra! Oh Cat! These are vocatives. Can one address the ego, Freud notwithstanding, in the same terms: Oh I!? [This broke me up when I first read it.]

Do you know that Vergillius Maro Grammaticus invented twelve different kinds of Latin, with a grammar for each, and forged 'classical' verses to exemplify them? See if you can find his De laudibus indefunctorum *(In praise of those not dead).*

Are you up on St. Anthony the Fool (ὁσαλόσ), a marvelous early Byzantine fellow (life edited, as I recall by the Bollandists) who traveled around the villages of the East with a girl friend as a prototype of a Kerouac hero? [Bollandists: the group of editors of the *Acta Sanctorum* from the 17th century to the present.]

Perhaps, after your study of these gamy fellows, you will think that the Mormon patriarchs in their original, unsandpapered versions are pale Galileans.

Vale, David

No. I didn't know about Virgilius Maro Grammaticus; I would try to find out. This remained a project for the future. I've been told that I have plenty of ego, but I still haven't learned to address it in the vocative case.

Fredegund

"Look *into Fredegund,"* David told me. *"Read what Saint Gregory of Tours wrote about her in the* Decem Libri Historiarum. *The two were at daggers drawn."*

Well, I checked her out. In the roster of wicked ladies she is at an all-time high and deserves to be cited in *Guinness World Records.* Fredegund (d. 597) turned out to be the second wife of the Merovingian King Chilperic, who ruled in northern France.

Chilperic had a number of sons. Fredegund wanted her own children to succeed—don't we all?—but in her drive to this end, she had all possible contenders within the

Egged on by his concubine Fredegund, Chilperic strangling his wife Audovera

Fredegund by Praetextatus's death bed

various kinships wiped out. She had one stepson, Clovis, murdered; another stepson, Merovech, she drove to suicide. Praetextatus, the bishop who was Merovech's baptismal sponsor and who was soft on him, she put to trial, exiled, and ultimately had him done in.

The Fabricated Letters of Antony and Cleopatra

D avid loved to talk about forgeries. One day in his office, he gave me a photocopy of a few pages of a lengthy document whose title page I've reproduced at the bottom of page 80.

PD: *What's this?*

DP: *This is an edition of Petronius Arbiter's* Satyricon *that contains faked letters between Cleopatra and various men. Amsterdam, 1669.*

I flipped through the pages. All in Latin.

PD: *I see it's in Latin. What am I supposed to do with it?*

I've had little Latin and less Greek. I'm on par with Shakespeare.

DP: *Well, you'll understand a few words here and there. And if Cleopatra interests you, maybe it'll spur you on to learn a bit more Latin.*

Later, back in my own office, I read the section that begins on page 148. It opens with an ode. And then comes

DE PRIAPISMO

SEV

Propvdiosa Libidine

CLEOPATRÆ REGINÆ

Ejusque Remediis

EPISTOLÆ

In lucum prelatae ex Bibliotheca

MELCHIORIS HAMINSFELDII

GOLDASTI

What remedies? The library of Goldast? Who is he? Well, I never used the text to improve my Latin, but I did check up on this Melchior Goldast (1576–1635) from Haminsfeld and what he put into his version of Petronius. It turns out that his book is really all about cooking. Yes, cooking! You may recall that the *Satyricon* of Petronius has a long and famous section

Title page of the 1669 edition of the *Satyricon*

on the Banquet of Trimalchio, the banquet to end all banquets. So what Goldast (who was, apparently, quite a scholar of classics, and incidentally a book thief) did was to compose a series of letters between Cleopatra, Mark Anthony, and a certain physician named Quintus Soranus.

Mark Anthony, his energies exhausted by the exertions of generalship and a very long night with Cleo, asked Soranus how to tamp down her hot libido. Soranus recommended for her an anti-aphrodisiac diet of rough

Newly found coin that belies Cleo's famed beauty

bread, lettuce dressed with vinegar and salt, a bit of meat, and wine of poor quality. For Anthony, he prescribed radishes (a homeopathic remedy, I suppose, instead of Viagra).

Regarding further details, I recommend my readers get a copy of Goldast's amplified edition of Petronius for the denouement.

Elagabalus

I f the very model of a modern major general knew all about the teenage emperor of Rome and an absolute monster, then we have a duty to keep pace. To set the scene gradually, I'll begin with a famous quote from Chapter 3 of Edward Gibbon's *Decline and Fall of the Roman Empire*.

> If a man were called to fix the period in the history of the world, during
> which the condition of the human race was most happy and prosperous,
> he would, without hesitation, name that which elapsed from the death of
> Domitian to the accession of Commodus.

More than 200 years have passed since these words were written, and I wonder whether current historians would alter Gibbon's judgment. Domitian reigned from 81 to 96. Commodus reigned from 177 to 192. In between,

Marcus Aurelius

we have Nerva, Trajan, Hadrian, Antoninus Pius, Marcus Aurelius, and Lucius Verus. The first five in this list have often been called "the five good emperors."

Let me pause a moment and ruminate just a bit about Marcus Aurelius (121–180) because he is connected with a David story. On the campus of Brown University in Providence, Rhode Island, there is a copy of a famous equestrian statue of Marcus Aurelius. The statue is located on a crest of a small hill just behind Sayles Hall. It had been there for many years and its surface deteriorated a bit.

Finally, about 15 years ago, Brown got some money for its restoration. The statue was hauled down, sent off to a restorer, and some months later came back in pristine condition.

In good weather, students sit on the grass and read books or (these days) work on their laptops; they rarely look up at Marcus, and I doubt whether many of them even know or care who Marcus was. Near the base of the statue there are a few benches, and on my way to David's office I used to sit there for a few moments and think about the statue.

There he sits on his horse, looking into the far distance with his right hand outstretched. I like to imagine him on a hill in the outskirts of Vienna (Vindobono, where there was a Roman Legion outpost) facing the east, pointing to the borders of his empire under attack on the far side of the Danube River. He faces the necessity of defending them with his legions. At its maximum extent in 116 the Roman Empire reached from Scotland to Iraq. It was now fraying seriously at the edges.

Marcus knew this and approached the matter philosophically. After all, he was not only emperor but also the author of *Meditations*, a book of stoic philosophy. Sitting there before Marcus I had my own meditations, and they led me to think that here, in this statue, if properly approached, could lie an extraordinary liberal education—bar none. One could put together a course centered around the age of Marcus that embraced the history, philosophy, science, mathematics, theology, arts, literature, etc., of the age.

Let's get a few faculty enthusiasts together, I suggested to David, and get going on it.
I volunteered to handle mathematics (which
in Marcus' day was in a state of decay). David
said he would join in and why didn't I start
the ball rolling. Weeks passed; months passed.
Other commitments interfered, and my Marcus Aurelius project became only a passing
thought.

Heliogabalus

And now from the good to the monstrous.
Moving forward from Commodus come Perti-
nax, Didius Julianus, Septimius Severus, Caracalla, Geta, Macrinus, Diadu-
menian, and finally we reach Elagabalus (a.k.a. Heliogabalus), a teenager who
sat on the throne from 218 to 222.

David introduced Elagabalus to me in the following letter:

Box 1900
25 March, 1980

*Dear Phil: You should really try to get a hold of a translation (I know there's at
least a French one) of Aelius Lampridius' Life of Heliogabalus in the Historia
Augusta. It starts charmingly:*

*"I would never have committed the life of Heliogabalus Antoninus (who was
also called Varus) to writing in fear that someone might learn that he was an Emperor of the Romans if that same Empire had not previously had Caligulas and
Neros and Vitelliuses. But, just as the selfsame earth bears not only poisons but
also grains and other helpful things, not only serpents but flocks as well, so the
thoughtful reader may find himself some consolation for these monstrous tyrants
by reading of Augustus, Trajan, Vespasian, Hadrian, (Antoninus) Pius, Titus, and
Marcus (Aurelian). . . ."*

*Lampridius then proceeds to the details of Elagabalus' dubious conception (his
name Varus was said to have been given to him in honor of the variety of his po-
tential paters). But Heliogabalus was not only noted for his perversity (a memory
of which is found in a horoscope published in POxy 46—no. 3298; cf. also 3299),
but for his clothes and furniture, his botanical and zoological collections, and his
patronage of actors, barbers, artisans, astrologers, and musicians.*

*I suggest and recommend him particularly because he is one of the few authenti-
cated cases I know of someone who sincerely worshipped himself. An Indian embassy was also sent to congratulate him on his election to the βασιλεία (the imperium);*

the embassy instructed Bar Daisan in Indian symbolism, whose account, as retold by Porphyry, is preserved in the learned Stobaeus. A splendid fellow.

I will quote later parts of this letter when I get to the cases of Apollonius of Tyana and al-Biruni. Brushing aside the question of who Aelius Lampridius (fourth century) was (not to mention Bar Daisan, Porphyry, and Stobaeus as well), let's get to the poisoned snake stuff, the perverse stuff—which some critics have said was merely pornography laid on thick. I rely on an English version of Lampridius in the Loeb Classics, translated and with notes by David Magie. It's important to realize Elagabalus came to the throne at the age of 14 and lasted only until he was 18 and that his story is bound up with the nature of Greek and Roman homosexuality. I quote from Lampridius:

> The Emperor was a priest of Elagabalus, (in Aramaic, El-Gebal) the sungod of Emesa on the Orontes River, and who was worshipped in the form of a black conical stone that was reputed to have fallen from heaven. Accordingly, as Emperor, Elagabalus tried to have the sun-god Elagabalus installed in Rome as the chief god of the Empire, closing down traditional temples and rites and attempting to extinguish the Eternal Flame. . . . He also celebrated the rite of Salambo with all the wailing and the frenzy of the Syrian cult. In fact, he asserted that all gods were merely the servants of his god, calling some its chamberlains, others its slaves, and others its attendants for diverse purposes. And he planned to carry off from their respective temples the stones which are said to be divine, among them the emblem of Diana, from its holy place at Laodicea, where it had been dedicated by Orestes. . . .

> He violated the chastity of a Vestal Virgin, Aquilia Severa, but after he divorced his first wife Paula, he married her. He removed holy shrines and thus he profaned the sacred rites of the Roman nation. . . .

All these things were serious no-nos and thereby Elagabalus foreshadowed his own doom. The early Christians were much better at extirpating the established rites and sites. But back to Aelius Lampridius:

> Elagabalus lived in a depraved manner and indulging in unnatural vice with men. . . . The soldiers who essentially had voted him in began to have second thoughts. Once established in Rome he sent his agents out to find men to search for those who had particularly large organs and bring them to the palace in order that he might enjoy their vigor. He used to have

the story of Paris played in his house, and he himself would take the role of Venus, and suddenly drop his clothing to the ground and fall naked on his knees, one hand on his breast, the other before his private parts, his buttocks projecting meanwhile and thrust back in front of his partner in depravity. He would likewise model the expression of his face on that with which Venus is usually painted, and he had his whole body depilated, deeming it to be the chief enjoyment of life to appear fit and worthy to arouse the lusts of the greatest number. . . .

Elagabalus also sacrificed human victims, and for this purpose he collected from the whole of Italy children of noble birth and beautiful appearance, whose fathers and mothers were alive, intending, I suppose, that the sorrow, if suffered by two parents, should be all the greater. Finally, he kept about him every kind of magician and had them perform daily sacrifices, himself urging them on and giving thanks to the gods because he found them to be well-disposed to these men; and all the while he would examine the children's vitals and torture the victims after the manner of his own native rites.

At his banquets Elagabalus would distribute chances inscribed on spoons, the chance of one person reading "ten camels," of another "ten flies," of another "ten pounds of gold," of another "ten pounds of lead," of another "ten ostriches," of another "ten hens-eggs," so that they were chances indeed and men tried their luck.

These he also gave at his games, distributing chances for ten bears or ten dormice, ten lettuces or ten pounds of gold. Indeed he was the first to introduce this practice of giving chances, which we still maintain. And the performers too he invited to what really were chances, giving as prizes a dead dog or a pound of beef, or else a hundred aurei, or a hundred pieces of silver, or a hundred coppers, and so on. All this so pleased the populace that after each occasion they rejoiced that he was emperor.

(I agree with Lampridius; this kind of thing is still around.)

In his letter to me, David mentioned actors and barbers; so before ringing down the curtain on this monster, let's hear about them:

During his reign Zoticus [whose father had been a cook] had such influence that all the chiefs of the palace-departments treated him as their master's consort. This same Zoticus, furthermore, was the kind to abuse such a degree of intimacy, for under false pretences he sold all Elagabalus' promises and favors, and so, as far as he could, he amassed enormous wealth. To some men he held out threats, and to others promises, lying to them all, and as he came out from the Emperor's presence, he would go up to each and say, "In regard to you I said this; in regard to you I was told that," and "In regard to you this action will be taken." That is the way of men of this kind, for, once admitted to too close an intimacy with a ruler, they sell information concerning his intentions, whether he be good or bad, and so, through the stupidity or the innocence of an emperor who does not detect their intrigues, batten on the shameless hawking of rumors.

This sounds very familiar to me in today's context.

With this man Elagabalus went through a nuptial ceremony and consummated a marriage, even having a bridal-matron and exclaiming, "Go to work, Cook"—and this at a time when Zoticus was ill. After that he would ask philosophers and even men of the greatest dignity whether they, in their youth, had ever experienced what he was experiencing—all without the slightest shame. For indeed he never refrained from filthy conversation and would make indecent signs with his fingers and would show no regard for decency even in public gatherings or in the hearing of the people. . . .

As prefect of the guard Elagabalus appointed a dancer who had been on the stage at Rome; as prefect of the watch he appointed a chariot-driver named Cordius, and as prefect of the grain-supply a barber named Claudius and to the other posts of distinction he advanced men whose sole recommendation was the enormous size of their privates. As collector of the five-per-cent tax on inheritances he appointed a mule driver, a courier, a cook, and a locksmith. He delighted in touching or fondling them, and whenever he drank one of them was usually selected to hand him the cup.

Enormities after enormities piled up—what I have reproduced here is only the tip of the iceberg. Ultimately, the soldiers of the guard

fell upon Elagabalus himself and slew him in a latrine in which he had taken refuge. Then his body was dragged through the streets, and the soldiers further insulted it by thrusting it into a sewer. But since the sewer chanced to be too small to admit the corpse, they attached a weight to it to keep it from floating, and hurled it from the Aemilian Bridge into the Tiber, in order that it might never be buried. The body was also dragged around the Circus before it was thrown into the Tiber.

But evil men can accomplish nothing against the upright. For no power could induce any to commit so great a crime, and the weapons which he was making ready for others were turned against himself.

Well, if you can believe this moralizing of Lampridius, you can believe anything at all.

A Few Thoughts on Mathematics and Theology

After a while, running down the references that David provided me became an addictive occupation, and, as you will see shortly, slightly obsessive. One noon, as we were having lunch together in the Blue Room cafeteria in Faunce House (he with his usual tomato and cheese sandwich and I with my bowl of chili) he asked me whether I still wanted a few more characters to fill my bedtime reading.

PD: *Yes, I've been running on near empty.*

DP: *Then why don't you look into Napier?*

PD: *Napier? The Scotsman who invented logarithms?*

DP: *The same.*

PD: *What makes Napier so interesting? I only know him for the Naperian logarithms. I thought that mathematicians, on average, were a dull lot—I mean, away from their specialties.*

DP: *Well, they say that Napier performed black magic with the aid of a black cock.*

PD: *The cock was his "familiar," I suppose. Anything else?*

DP: *Well, this much is certain: he used numerology to predict the date of the Apocalypse.*

That was enough to get me going. It's a long story that developed out of this lunch. I took on the persona of a Sherlock Holmes of history and I'll tell about my investigations in some detail.

I have long been interested in connections over the centuries between mathematics and theology, and at one point I was invited to lecture on the subject. My old student Rick Vitale led me to a friendship with Al White at Harvey Mudd College in Claremont, California, and Al's interests in "humanistic mathematics" got me an invitation in the spring of 1998 to be one of the speakers at a several-month Scripps College program entitled Fin-de-Siècle Soul: The Status of Non-Material Being in the Late 20th Century.

Wow! Most of what I know about the soul can't be verbalized. I asked my hostess who some of the other speakers were in the long program, and I remember the name of John Cleese, of *Monty Python* and *Fawlty Towers* fame.

Wow again! Apparently, Cleese is not only a very funny fellow but also a quite scholarly one. I got nervous. The topic I selected, Mathematics and God, would be very heavy cake indeed, and though commensurate with the grandiosity of Fin-de-Siècle Soul, Cleese would be a very hard act to follow. My talk would sink much as the battleship *Missouri* in years gone by sank in the mud of Chesapeake Bay.

Why had I selected this topic, considering that I am neither a historian of mathematics nor a theologian? Simply that for some time I had a bug in my ear and here is the buzzing thereof.

We are living in a mathematical age; our lives, from the personal to the communal, from the biological and physical to the economic and even to the ethical, are increasingly mathematized. This is the case even if the average person has little necessity to deal with the mathematics on a conscious level. Mathematics permeates our world, often in "chipified" form. According to some theologies, God also permeates our world; God is its origin, its ultimate power, and its ultimate reason. Therefore it makes sense to inquire what, if anything, is the perceived relationship between mathematics and God, and how, over the millennia, this perception has changed, and what its consequences are.

I spread the word quietly among my mathematical friends that I was going to lecture on mathematics and theology; I wanted to get a reaction and perhaps even a suggestion or two. One research mathematician, who in his personal life would be considered very devout in a traditional religious sense, told me that "God could never get tenure in my department." Another friend, a European mathematician, well versed in the history of mathematics, told me that "The relation between God and mathematics doesn't interest me."

I think that these two reactions sum up fairly well the attitude of today's professional mathematicians. Though both God and mathematics are everywhere, mathematicians tend towards agnosticism; or, if religion happens to play a role in their personal lives, it is kept in a separate compartment and seems not to be a source of professional inspiration.

But this has not always been the case. And this fact is not appreciated or is widely disregarded. The number of books dealing with the relation between science and religion is enormous, particularly in the post-Darwin period. In these books you will find discussions of conflict, agreement, independence, dependence, accommodation, reconciliation, apologetics, hermeneutics (i.e., modes of interpretation, particularly of sacred texts), etc.

These two themes, science and theology, constitute what the late evolutionist Stephen Jay Gould referred to as "two magisteria," and Gould suggested that the two had little business in overlapping and messing with each other; they ought to live side by side peacefully in a non-aggression pact. But, alas—or gratefully, depending on one's point of view—these two themes do overlap and always will.

To my knowledge (and I've checked this out with several historians of science) there is no book that deals in depth with the 4,000-year history of the relationship between mathematics and God. There are numerous articles and books that deal with particular chapters of the story: Joan Richard's *Mathematical Visions: The Pursuit of Geometry in Victorian England* comes to mind. It describes the conflict of the discovery of non-Euclidean geometry with the theology of the period. But most historians of mathematics in the past two centuries, under the influence of the Enlightenment and of positivistic philosophies, have avoided the topic like the plague.

This neglect—probably in the name of purity or the avoidance of triviality—is an act of intellectual cleansing that parallels the many acts of iconoclastic destruction that have overtaken civilization at various times and places. Why has it occurred? Numerous reasons have been suggested, but occurred it certainly has.

The extent of the relationship between mathematics and theology should not be underestimated. There is much that can be, and has been, said.

Practically every major theme of mathematics—its concepts, methodology, philosophy, number, geometry, pattern, computation, axiomatization, logic, deduction, proof, existence, uniqueness, non-contradiction, infinity, randomness, chaos, entropy, fractals, self-reference, catastrophe theory, description, modeling, prediction, determinism, evolution, omnipotence, omniscience—have been linked by someone somewhere and in some way to theological concepts. As examples: does God have the power to make 2 + 2 other than 4? Does God know what lies ahead for the cosmos?

These links are part of the history of mathematics. They are a part of the mathematical civilization into which we were born. They are part of the applications of mathematics. In recent years they have been extended to embrace theological links to cognition, personhood, feminism, ethnicity, etc., developed along mathematical lines.

The contributions of mathematics to theology have been substantial; in the other direction the contributions of theology to mathematics are arguably less substantial. The young John Newman (later, Cardinal) argued that the statements of mathematics were more firm than those of dogmatic theology. Hermann Cohen (1842–1918), the philosopher, thought that mathematics was the basis on which theology must be built. In recent discussions, as we shall see, mathematics takes intellectual priority over theology just as it did for Cardinal Newman. By way of contrast, one should remind oneself of

Hermann Cohen

the hierarchical order in the days of the Scholastics (e.g., Thomas Aquinas): mathematics, philosophy, and metaphysics, with theology at the apex.

Despite the opinions that point in one direction—from mathematics to theology—it is by no means clear to me that theology has contributed little to mathematics. For example, the demands of religious ritual and not religious dogma raised questions for mathematics, and this pressure led to improved techniques and theories. Thus, both the church and secular calendars are mathematical arrangements. The Jewish-Arabic philosopher and theologian Moses Maimonides (1135–1204) wrote a book entitled *On the Computation of the New Moon*. The demands of astrology for exact planetary positions, demands that very often had links to theology particularly in eastern religions, stimulated and supported mathematics for long periods of time. Such demands have certainly been much less publicized than, for example, the demands of the military.

Claims have been made and descriptions have been given of the manner in which Christian theology entered into the development of Western science. Here is the view of physicist Freeman Dyson:

> Western science grew out of Christian theology. It is probably not an accident that modern science grew explosively in Christian Europe and left the rest of the world behind. A thousand years of theological disputes nurtured the habit of analytical thinking that could be applied to the analysis of natural phenomena. On the other hand, the close historical relations between theology and science have caused conflicts between science and Christianity that do not exist between science and other religions. . . . The common root of modern science and Christian theology was Greek philosophy.

The same can be asserted of mathematics, though perhaps with somewhat less strength.

A few Western opinions over the ages, arranged more or less chronologically, should give us the flavor, if not the details, of the relationship between mathematics and theology. I have less knowledge of non-Western writings (e.g., the Oriental or the Indian). In the latter, as David once assured me, the necessities of precise numerological altar construction as described in the Sulba-Sutras c. 450 BCE led to the development of mathematical techniques.

While quoting and citing is easy, what is not easy is to enter into the frame of mind of the authors quoted and of the civilizations of which they were part, and how the particular way they expressed themselves mathematically entered into the whole. It helps to remember that the secularization and the disenchantment (i.e., the "demagicization") of the world are relatively recent events to be located, say, in the late 17th century.

To reiterate the bug in my ear: the neglect of non-rational elements in the development of mathematics—probably in the name of purity or to avoid triviality—has been an act of intellectual cleansing, and the story of Napier looked promising.

John Napier
The Master of the Black Cock

A t this point in typing my story, I was interrupted by receiving an e-message from Moscow. I had been pursued by the head of one of the laboratories of the Russian Academy of Sciences who seriously questioned something I wrote in one of my *SIAM News* columns (i.e., in the newsletter of the Society for Industrial and Applied Mathematics, October 1999).

> . . . the scientific method was simply to try everything. To which quantum physicist David Park appended "but don't waste time being stupid."

This maverick view is pretty close to the view of Paul Feyerabend, who was much more of a philosopher of science than I am. My Moscow correspondent then produced a 400-word summary of his own explanation of what the scientific method was. I acknowledged its receipt politely and forgot about the matter completely. So back to my story.

Turn around, they say, is fair play. Readers chase after me, and occasionally I, as a reader, chase after an author questioning a statement he or she has made. This doesn't happen very often, but it does happen. I have heard it stated by some puritans that a reviewer has a moral obligation to point out to a writer where and why he/she is wrong. I do it on a very, very selective basis. During the morning of the particular day on which I picked up a book by Eugen Weber, I had noticed something peculiar about myself. Was it an omen?

There are just two ways in which a belt can be threaded through the belt loops of a pair of pants. In the first, the end of the belt, usually tapered in some fashion, points to the right. In the second, it points to the left. I suspect that most people stick to one way throughout their lives. I am a right-handed belt-looper, but on that particular morning I noticed that I had been wearing my belt with the left orientation. This is going to be a remarkable day, I said to myself, and so it was, if one admits the elevation of one's eyebrows into the canon of remarkable events.

The event occurred early in 1999. The millennium was upon us. The Y2K (i.e., the year 2000) fears about errors in computer programs were blown out of proportion by the media. The amount of millennium and apocalyptic material grew both scholarly and crazy, both from scientists and from non-scientists, both from clerics and the laity, both in print and on the Web. The world famous physicist Stephen Hawking worried about the greenhouse effect. He made his computation, and said that he feared the human race would not survive another millennium. It was hard to avoid this kind of material.

Taking advantage of the interest raised by the calendric switch to the 21st century, a distinguished historian at UCLA, Eugen Weber (there is a UCLA chair in history now named after him), put out a book entitled *Apocalypses: Prophesies, Cults, and Millennia Beliefs Through the Ages.*

I picked up a copy, just hot off the press, and enjoyed Weber's treatment. One sentence on page 92, though, raised my eyebrows:

> Napier said to have valued them [i.e., logarithms] because they speeded up his calculations of the Number of the Beast.

Before getting on with my story, I should explain four things: who Napier was, what logarithms are, what the Number of the Beast is, and why my eyebrows shot up. I'll start with the item on this list that is the best known: the Number of the Beast.

In the book of Revelation (the last book of the New Testament), we read in 13:16–18:

> Here is wisdom. Let he who has understanding calculate the number of the Beast; for his number is that of a man and his number is 666.

The Beast of Revelation

(There are, of course, many translations from the original Greek into English, some of them explaining that in those days numbers were often represented by letters, so that person's names—indeed, all kinds of words—have numerical equivalents.)

The book of Revelation is essentially a vision of the future elicited by the feeling that Rome and its government have been monstrous. Dire happenings are predicted and these, after some time, are followed by the Messianic Era, the Second Coming, the New Jerusalem, and, effectively, the end of the world. Numbers abound in Revelation: the 24 elders, the seven lamps, the book with seven seals, the seven vials, the 144,000 Israelites who received the seal, the four horsemen of the Apocalypse, the seven thunders . . . and on and on.

Over the millennia, Revelation has appealed powerfully to mystics, Kabalists, those inclined to numerology, to the apocalyptic, millennialists, and to messianic spirits. The amount of commentary on Revelation as well as the number of interpretations is immense. The number and variety of calculations that refer to 666 and that go back and forth between numbers and letters in Latin, Greek, and Hebrew, are beyond imagining.

Reader, you probably know that 2/3 = .666666666 . . . , but did you know that $1 + 2 + 3 + 4 + \ldots + 36 = 666$? And $36 = 6 \times 6$. Did you know that the squares of the first seven prime numbers add up to 666:

$$2^2 + 3^2 + 5^2 + 7^2 + 11^2 + 13^2 + 17^2 = 666.$$

And isn't it also the case that the prime number 7 is a magic or holy number? Did you know that the fourth root of π ($\pi^{1/4}$) is 1.331 335 363 . . . and that $331 + 335 = 666$?

If these facts thrill you, if they send a frisson up your spine, then you are a bit of a numerologist, a neo-Pythagorean. If you attribute arcane or transcendental potency to individual numbers, if you read into these arithmetic relations more than mere arithmetic, more than that coincidences favor the prepared mind, then you, as well as thousands of others, have a numerological gene. Not to worry. You are in good company: modern cosmologists have such genes. And so did John Napier.

Napier (1550–1617) is known primarily as the discoverer or inventor of logarithms. Now, if we read his biographers, and the principal source of information is *Memoirs of John Napier of Merchiston* (London, 1843) by his descendant Mark Napier, Napier's intellectual life seems to be split down the middle. On the one hand, he was a mathematician and the designer (in his imagination) of lethal war engines for "the defense of the island against the threatened Popish invasion from Spain." On the other hand, Napier was one of the principal apocalyptists of his age, a fierce Protestant, violently anti-Rome, a man who "proved" that the pope was

John Napier, baron of
Merchiston (Scotland)

the Antichrist, and a man who computed the date of the end of the world. Chronologically, his apocalyptic thoughts set out in *A Plaine Discovery of the Whole Revelation of Saint John* (1593) predate his *Mirifici Canonis Descriptio Logarithmorum* (1614). There is some evidence that Napier considered his theology his prime achievement and his mathematics a mere hobby.

Along with these two interests and achievements, there is also the fact that

he possessed the respect and confidence of the most able and Christian pastors of the Reformed Church, and while he was looked up to and consulted by the General Assembly, of which he was for years a member, he was at the same time regarded . . . as one who possessed certain powers of darkness, the very character of which was in those days dangerous to the possessor. (*Memoirs of John Napier of Merchiston*, 214)

In short, "It was firmly believed, and currently reported, that he was in compact with the Devil." Rumor had it that Napier performed acts of black magic with the aid of a black cock. Rumor had it that he made a very strange contract with a highway robber. What probably saved him from the stake was his high reputation as a theologian.

Enough for Napier the man—now for his logarithms. I hardly have to explain to my contemporaries what they are, for every high-school algebra book had a chapter on them and a well-thumbed table at the back. My high-school math teacher had us drill on logarithms to the point of boredom. Since addition is a relatively easy operation and multiplication is difficult, logarithms were conceived of as a mathematical device for reducing complicated multiplications and root extractions (powers) to simple addition. They outlived this function years ago with the invention of mechanical computation machines, and then the invention of the electronic digital computer rendered even the mechanical devices obsolete.

Nonetheless, the logarithm as a mathematical function (i.e., a graph, a curve, etc.) and its inverse, the exponential function, are completely alive and well, operating as two of the absolutely basic and fundamental tools of pure and applied mathematics.

In an inaugural address (July 1914) by Lord Moulton at the Napier Tercentenary Celebration, Moulton said logarithms were

a bolt from the blue; nothing had led to it, foreshadowed it, or heralded its arrival.

Without in the least detracting from Napier's accomplishment—the step he took was considerable—the previous sentence is patriotic nonsense. Historic surveys published in the *Napier Tercentenary Memorial Volume* provide an accurate and adequate description of the mathematical predecessors of logarithms.

I must now explain why my eyebrows shot up when I read the sentence in Weber's book. I was well aware that Napier, the inventor of logarithms, was a student of the book of Revelation. David had let me in on this. I knew that in his day, science and theology often resided in one and the same mind. I was aware, for example, that the great Isaac Newton, a half century later, wrote speculations on the book of Daniel, and considered that work to be his magnum opus. Were these two seemingly diverse subjects—diverse from the understanding of our 21st century—were they resident in two separate boxes of the mind or did the boxes intercommunicate? My view is that they did, but that it would be very difficult to establish how they did.

I walked across the Brown campus and asked David whether Napier's apocalypticism fed into his discovery of logarithms. Pingree said he never heard the story. I said tentatively, "Se no e vero, e ben trovato." (If it ain't true, it oughta be.) Pingree agreed mildly to that.

"*You know,*" David went on, "*astrology, numerology, Kabala, were all part of the scientific or pre-scientific air. It was in Napier's family even. His cousin, Dr. Richard Napier, and his cousin's son, Sir Richard Napier, were both students of the* Picatrix."

"What's the Picatrix?"

"*A book about Iranian Astro-Magic. Old, old stuff. Weird.*"

Leaving David's office quite by accident, I bumped into Noel Swerdlow, just in from the University of Chicago. Swerdlow is an expert in the history of astronomy. I asked him about Napier. Swerdlow answered that he never heard the story and in any case he didn't believe it. Besides,

NS: *You don't need logarithms to compute the apocalypse. Addition, subtraction, a touch of easy multiplication should do it. The story must have been spread by mathematical ignoramuses.*

PD: *But what if Napier had devised a complicated system? Epicycles, eclipses. World-shaking consiliences? The golden number. The orientation of the stones in Stonehenge; Lord knows what else?*

NS: *Very, very unlikely.*

PD: *You know that today's cosmologists are computing the date of the apocalypse—in modern terms, of course. The mathematics is non-trivial. Comet orbits. That sort of thing. Need computers.*

NS: *Yes, and they compute backward to the big bang. I'm in the astronomy department. Those fellows come up with new theories—models they call them now—every Monday and Thursday.*

PD: *Do they go forward in time?*

NS: *Sure. Proton decay. One proton lost every zillion years.*

Then Noel rushed off: he was on a visiting committee.

• • • • • • • • • • •

Some weeks later, I ran into Gerald Toomer.

PD: *Gerry, is it possible that Napier developed logarithms so he could compute the date of the apocalypse easier? You know he was a great apocalypticist.*

GT: *I know nothing about Napier. But I know this: that in the absence of some kind of documentation, it is very difficult to establish peoples' motivations. Do people know their own motivations, often unconscious?*

Is it possible, Gerald Toomer asked me as we continued our conversation, whether 400 years from now, all our present cosmology, black holes, big bangs, etc. will appear as ridiculous to the scientists of the future as Napier's computations on the Apocalypse do now to us. (A few pages further, I'll display some of these computations.)

I wanted to pursue the two-box idea, and whether consciously or unconsciously, there is spillover from one box to the other. I made a luncheon date with Prof. Jim Anderson, neural scientist and student of cognitive and linguistic sciences, about the possible carry over from one intense interest to another, which to us seem totally disconnected.

JA: *Interesting question. We don't know the answer. There are many cases of apparent disconnections. The motivations are hard to pinpoint. The brain can be a surprising thing. Sometimes physical abnormalities can play a role. Substances can play a role. Did Newton poison himself with mercury as has been rumored?*

PD: *I never heard that story.*

JA: *Well, you might look into that aspect.*

Though fascinating, I had no intention at the moment of going off on that particular tangent.

Undaunted by all these opinions, I wrote to Professor Weber asking him on what authority he composed that sentence about Revelation and logarithms. Weber answered me very kindly and thoughtfully, and said I should look at a previous book, *Century's End* (Doubleday, 1990) by Hillel Schwartz, a Yale PhD and a cultural historian.

An immediate trip to the library produced Hillel Schwartz's book. Schwartz makes reference to about 20 pages of *The Apocalyptic Tradition in Reformation Britain: 1530-1645* by Katharine Firth.

Reading Firth's book as carefully as I was able—the book derived from an Oxford University doctoral thesis under Hugh Trevor-Roper—I found no direct evidence for Weber's statement. We know that Napier computed the date of the Apocalypse. Here, for the flavor, and as a verification of Swerdlow's views on the mathematics that Napier found necessary for the computation of the Apocalypse, is a clip from Katharine Firth's book, pages 139–145. The terms "seals," "vials," "trumpets," and "angels," are what Firth calls the symbols in the book of Revelation.

> In order to decide the most likely candidate for each image or symbol, Napier looked first for the natural divisions of the text. In the first treatise [i.e., Napier's *A Plaine Discovery*] he discovered three stages of history. In the first, the prophesies referred to the time of the baptism of Christ to the destruction of the Temple of Jerusalem, from AD 29 to 71, and were hidden under the signs of the first six seals. In the second, history from the opening of the seventh seal (AD 71) to 1541 was hidden under the terms of the seven trumpets and seven vials, which he showed to be concurrent.
>
> The proof of the concurrence of trumpets and vials was central to his argument. This gave him his first date. The fifth trumpet spoke of a star that fell, and this he identified as Muhammad, a prophet "who fell from his former Christian profession and became an apostate." This fifth vial spoke of a plague of locusts that would ravage the earth for five months. This referred to the consequent rise of the heretical Turks. The time of the sounding of the fifth trumpet and the pouring-out of the fifth vial he set at 1051, or as he said at about the time of the domination of the Turks under Zadok.
>
> Napier then set forth a few numerical propositions to prove that great alterations of kingdoms took place roughly every 245 years. He assumed that as the Hebrews had measured history by Jubilees, periods of 49 years, so ordinarily should historical periods be measured. This law, however, had been altered by Daniel's prophecy to a period of seventy weeks or 490 years, as one day was generally taken to signify one year. Alterations could then be thought to take place every 490 years. Napier proposed this correction to Carion [John Carion, a mathematician, astronomer, and astrologer] and others who believed that the "fatal period of empires" was 500 years. The further revelation of John had again altered the periodization, showing Napier that the world would not last a full seven Great Jubilees, that is a full seven periods of 490 years each, or 3,430 years. To John it had been revealed, and by Napier discovered, that the periods would follow the seven trumpets and vials at intervals of one-half a Great Jubilee, or 245 years.

By working forwards and backwards from the fifth trumpet and vial of
1051, Napier followed the fortunes of Empire and recorded at approxi-
mately 245-year intervals the alterations proposed.

1. One trumpet and vial: the destruction of Jerusalem—71

2. The translation of the Empire to the east (Constantine and
    Pope Sylvester)—316

3. Totila burns Rome—561

4. Charlemagne made Emperor—806

5. Zadok Dominator of the Turks—1051

6. Osman—1296

7. Reformation—1541

The third division or the last age began after the sounding of the seventh
trumpet. This age appeared under the sign of the five thundering angels,
each angel governing a Jubilee, 49 years. The first three angels covered the
years from 1541 to 1639 to 1688. The fourth angel, "even Christ himself,"
was due in 1688. etc., etc.

Despite all this primary material, I was not entirely convinced by Swerd
low's assessment of the mathematics required. I recalled that the logarithms
Napier computed were not, as we understand the word now, the logs of the
numbers 1, 2, 3 . . . . They were the logarithms of the trigonometric functions.
And where, pray, might these be used? In navigation, in positional astronomy, and abstractly, in spherical trigonometry. Napier himself worked out
some convenient formulas in spherical trigonometry, still called by his name,
to facilitate computation there.

So the evidence would seem to be clear: what pushed Napier to invent
logarithms were the difficulties—the tediousness, if you will—of certain numerical calculations. But there was still a lingering doubt in my mind about
the disconnect between spherical trigonometry and the Apocalypse. For
what purpose were celestial computations wanted in those days? For navigation, of course. But there is another answer consonant with the thought and
activity of the scientists of Napier's day, and it was to predict planetary positions accurately, for the purposes of astrology.

I still had on the back burner of my word processor my partially written
document on the relation between mathematics and theology. I "pecked at it"

from time to time. Every once in a while I would find a sentence, somewhere, that would induce me to reopen the document labeled "Math & G" and peck a bit more.

For example, here is a clip from Lisbet Koerner's review of Thomas W. Laqueur's book *Linnaeus: Nature and Nation* that appeared in *The New Republic* on June 5, 2000.

> Laqueur's book is valuable not least for its subtle account of how our modern science developed not out of rejection of religion, but out of religion's core.

The question of the relation between John Napier's apocalypticism and his logarithms was still on my mind. Was this an instance of science developing out of religion's core, or wasn't it?

Katharine Firth

K atharine Firth's book seemed to me to leave the matter in an ambiguous position. I decided to get in touch with her and probe the question. Perhaps she knew more about the matter; perhaps she was in possession of more information than she had put in her book. What interested me didn't necessarily interest her.

But first of all I had to locate her. I asked some historians of my acquaintance but they knew nothing. I thought of writing Oxford University Press— surely they must maintain an address list of their authors. I delayed writing. A previous experience with this kind of inquiry had led nowhere; publishers can have very sloppy attitudes toward their authors, especially when a book is out of print and there is no necessity for paying royalties.

The world was then suffering from Y2K-itis, fueled by a few super twitchy computer mavens but driven largely by the media. Yes, there were problems, but they were such as could be overcome without subjecting the world to apocalyptic angst. Fighting fire with fire, I tried the Web; perhaps I might pick up something interesting under "apocalypse" or a variant expression.

I found hundreds of websites, but which one to visit?

One could spend hours, days, and weeks at this game. Was it luck, brains, or a mixture of the two that led me to *Frontline*, a TV show? It had recently done a show on apocalypticism and invited comments on its Website. Wild, wild stuff was aired there. But in the middle of the fantasies, irrationalities, and accumulated nonsense, I found several paragraphs, carefully reasoned and scholarly, in which the writer rapped the presentation. And at the end, the writer's signature was Katharine Firth—the author I'd been seeking!

> I do not think it insignificant (and I think the connection should have been made) that the rise of analytical approaches to philosophy, to all human knowledge including material scientific inquiry, could, when applied to literary texts including Scripture, be seen by the reformed and even some of the counter reformed, as pre-eminently literal and exact, scientific, calculable, logically persuasive from the axioms of the text and demonstrable from current events. This is the real origin of our western tradition as we encounter it today.

> Many highly respected and university trained minds, for example Isaac Newton, did what they thought was their best and most important work as interpreters of the book of Revelation, Daniel, and other passages. Similar expressions filter through the rationalism of the 18th century, the German idealists of the 19th and the Marxist, fascist et al, of the 20th centuries. Sharing aspects, language and image with all apocalyptic thought, the specific form we now have owes more to the last four centuries than to the era around the time of Christ. It would have been good to see this balance.

Here was gold among the dross; here was further material for my document "Mathematics and God." Additional Web-scrounging and I learned more. I could not find an e-address for Katharine, but I learned that she was living in a small village in central Maine. What on earth, thought I, was an Oxford PhD—who, I (mistakenly) assumed, was of English birth, whose thesis supervisor was the famous and notorious historian Hugh Trevor-Roper, later Lord Dacre—doing in a small village in the middle of Maine far from the principal centers of historical scholarship? Teaching, surely, in one of the nearby colleges: Bates or Colby, or perhaps the University of Maine at Farmington.

I made inquiries at these places. No luck. None of the historians or humanists there had heard of her. I tried telephone information. The mystery deepened. No number currently listed. Had she, as many other intellectuals had, succumbed to the minimal, ecologically correct life, living on fiddlehead ferns and blueberries in a hut adjacent to a large abandoned apple orchard? Had she moved back to England? Had she passed on?

I wrote to her care of general delivery at her village. And that worked. An e-connection was rapidly established and my bona fides rapidly accepted. I exchanged views and opinions, and after a few such back-and-forths, I ended with an invitation to visit her in her lakeside cottage.

Hadassah and I drove up the Maine Turnpike, stopping overnight in Lewiston. The whole trip, if done in one piece, would have taken us four and a half hours.

Katharine, once found, resided in a lakeside cottage she had named The Sparrows. Take a look at, she told us, Psalm 84:

> Even the sparrow finds a home
> And the swallow has her nest
> Happy are those O Lord of Hosts
> Who dwell in Thy house.

She was slight of build, mother of five, owner of a coon cat named Micah, sharp as a razor, gutsy, and the first woman, so she said, to be allowed into the Bodleian Library. A stimulating conversationalist, a bit unconventional, even a bit heretical, a trained scholar, a trained and practicing Episcopalian priest, and a delight to have met.

The story of her scholarly career that began brilliantly and was derailed, is briefly told. It is the classic tale (1970s) of prejudice against women, against Americans by the British, against the British by the Americans, against her field of expertise, against her overqualification. The sunny and the shady parts of her story are both remarkable and I encouraged her to write it down. She must tell it herself.

We had lunch at a lakeside restaurant: Caesar salad with anchovies for both ladies, a club sandwich for me. When I brought up the subject of the possible connection between Napier's apocalypticism and his logarithms, Katharine mentioned that at some point in history, there had been a fire that destroyed many of Napier's manuscripts. If we now had those sheets, the connection, if it existed, might have become apparent.

Katharine, who was a wicked mimic, told us about her experiences with Hugh Trevor-Roper, her thesis advisor. She used Oxford tonalities to reflect his initial snobbishness vis-à-vis her gender, her place of origin (Maine), her ethnic background (she was born a Kelley), and her undergraduate experience (Bates College). He was a man, she said, who was concerned totally with ideas. But evidently, despite his prejudices, when he was once confronted with the fact that she was a winner, a strong-willed and brilliant student who knew what she wanted to do, he took her on and gave her careful, painstaking, and constructive criticism of her project and of each sentence she produced.

I asked Katharine: "What about Trevor-Roper's boo-boo with regard to the Hitler diary?" Trevor-Roper considered, she said, the forensic evidence, physical and graphological, presented to him by the experts he consulted. He considered the historic and psychological evidence in which he was a world expert, and came up with a verdict. He said his business as a historian was to make such judgments. He did it on a daily basis and he had to have faith in his judgments, otherwise he could not function. What more, asked Katharine, can one ask of a historian? Anyway, after two weeks, when it emerged that one of the principals had lied to him, and when more evidence emerged, Trevor-Roper reversed his judgment, but there was a certain damage to his reputation.

This incident in Trevor-Roper's career put me in mind of Karl Popper's philosophy of science: total verification of a theory is difficult, if not impossible. What makes a statement scientific is the possibility of its disproof.

It also put me in mind of Attorney General Janet Reno's defense of the Department of Justice's action in a case of Wen Ho Lee and his alleged breach of atomic security at the Los Alamos Laboratory in September 2000. We consider, she said, the evidence that we have; we consider what the law says; and then we, weighing it all in the light of our own experience, make a decision whether or not to bring a charge.

We spoke briefly about Katharine's job at Grace Church in Bath. It was not all peaches and cream. She had made good friends there, but there were also those who did not like her theology. And she complained briefly about what has come to be known as the "stained glass ceiling." But most importantly, it was a temporary job and she would be out of work in a few months. I asked her if the opportunity presented itself, would she like to return to scholarly writing? She answered that she would. Did she have a project in mind? She did.

After lunch we said goodbye and I drove south, stopping overnight in Bethel, Maine, en route to Interstate 93 via Franconia Notch. We selected a bed and breakfast more or less at random from the tour book; a spacious house (we were the only guests that night) that had a collection of old books owned by an absentee landlord. I went to the shelves and selected a book at random: a volume of old detective stories by Marjorie Allingham. I opened the book to the first story, and found that the victim's last name was Dacre. Not exactly a common surname is it; Dacre is not a Smith or a Jones. Put it this way: there are no Dacres listed in the Providence phone book. Tell me, do coincidences chase after us or do we chase them?

In the days that followed our visit to Katharine, I "talked" on email with her about various academic possibilities that might enable her to resume her scholarly writing. With the help of a few academic colleagues, I located a fellowship designed for women who had been away from the life of scholarship for a number of years and wished to return to it. It seemed to me to be tailor-made for Katharine. The fellowship would require her presence in Cambridge, Massachusetts. It didn't pay much, and I asked her whether it would be feasible for her. She said it would and that she would work out an application.

And then, suddenly, on September 30, 2000, Katharine died in an automobile accident. A brief encounter, but should we measure the significance of a relationship by its duration?

I received a call from her brother, who lived down the road from her on the lakeshore. Near her word processor he had found the letter I'd sent her. His sister had died in an automobile accident, and he wanted me to know. I was confused, in tears. Shortly after that I had a call from one of the officials of Grace Episcopal Church in Bath, Maine, where Katharine was employed as a priest. Could I come to her funeral and present a short eulogy? I? A person who had hardly known her? I felt inadequate to the task. I was a stranger. I had intruded without invitation into a different part of her life, a part in which her family and congregants took little interest. But I, perhaps, was the only link left to Katharine as a historical scholar.

I have long harbored an idea—I'm not sure where I picked it up from— that at every rite of passage there is a mysterious stranger present who comes uninvited. The stranger is the chorus in Greek plays. The stranger represents the inexplicable course of the world, of human life. The stranger represents fate, but also comfort in the face of fate. Was I to be the stranger on this heart-rending occasion?

I could not face it. I told the church that I had a previous engagement (which was true enough) but that I would send them an appreciation of

Katharine, which they could use as they thought proper. I sent them a paragraph praising her character and scholarship. The parish administrator responded immediately that my words had caught Katharine's character exactly and that they would post them on a memorial billboard in the vestry.

In the weeks that followed, I probed and learned more about the accident. What I am about to tell comes largely from an article by Dennis Hoey, a writer for the *Portland Press Herald*, that was posted on the Web.

Katharine was born in Philadelphia in 1945. Her family moved to Maine, and she graduated from Portland High School and then went to Bates College in Lewiston. I have already said a few words about her career at Oxford. She married a young Englishman. In the '80s she worked for a time on the British Open University teaching 17th-century English history. She had five children. She was divorced. She returned to the States. She went to the Pittsburgh Theological Seminary.

Katharine was ordained as a priest in December 1989. In September 1999, she accepted a temporary position as rector at Grace Episcopal.

On the afternoon of September 30, 2000, she drove to Georgetown, Maine, a small village on the coast, and performed a wedding ceremony. After the ceremony, she attended the reception at the Grey Haven Inn. Later in the evening, she left the reception to drive home. She was unfamiliar with the area and she took a wrong turn that led her ultimately to a narrow opening and onto the 50-foot town pier. She drove off the dock, plunged into the water, and was drowned.

Several months later, the state's chief medical examiner released a report that at the time of the accident, Katharine's blood-alcohol level was twice the state's legal limit.

It is very easy to speculate and to write scenarios. In the weeks that followed my reading the news article in the *Portland Press Herald*, I often speculated, conjuring up in my mind sad and depressing possibilities. All "what if nots?" Ultimately, I said to myself: "Enough. I don't want to know more. I don't want to talk to those who were close to her and find out more. I want to remember her simply as the author of a book I read and admired. I want to remember her as a person who intrigued me."

I did initiate one inquiry. I wanted to get in touch with Prof. Hugh Trevor-Roper and ask him what he remembered about his old student. I asked some knowledgeable friends in England whether Trevor-Roper was still alive, and if so how I could reach him. They didn't know. I persisted. Ultimately, one

friend suggested that I write to him at the House of Lords. I did so, inform-
ing him merely that his old student had passed on, and asking him whether
he retained a few impressions of her. After a month an answer came. It was
typed on House of Lords stationery.

> *11 February, 2001*
>
> *Dear Prof. Davis,*
>
> *Thank you for your letter of 20 December about the late Katharine Kelley. I am very
> sorry to hear of her death, and must apologize for the delay in my reply.*
>
> *I remember Katharine Kelley well, but there is not a great deal that I can say
> about her. She was a model student, worked very well on her own, was punctual
> in producing material, and in the end produced an excellent thesis, which, as you
> know, was published by the Oxford University Press. It is in my opinion a very good
> book. In all her work she was quietly efficient, seemed to have no problems, gave no
> trouble. I had some correspondence with her afterwards when she was considering
> editing texts for a publisher. She had some difficulties in finding further employment
> in this country and I think then went back to America. But as I left Oxford in 1979,
> the date of my last correspondence with her, I have not seen her since.*
>
> *I am sorry to give you so jejune an account, but she was one of those students
> who, by her very virtue—by the fact that she gave no trouble—had only a slight
> impact. I do not seem to have had any letter from her since 1979.*
>
> *Yours sincerely,*
>
> *Hugh Trevor-Roper*

Life goes on and stories sad, cheerful, and strange accumulate.

Abu Ma'ashar and
The Hurrians

I t was not only through notes and letters that I learned about the inhabitants of David's Oddballs' Hall of Fame but also over many a lunch. I would ask David a question, perhaps by way of elucidating what he'd told me the previous week, and this would start him gushing out with a seemingly endless supply of material. Occasionally, out of his vast collection of articles and books (often his own writings), he would give me a few pages that he thought would bear on the characters in his Oddballs' Hall of Fame. Just for the flavor of the thing, I note the following articles in my personal collection of Davidiana.

- Kristin Lippincott and David Pingree, *Ibn-Al Hatim on the Talismans of the Lunar Mansions*

- David Pingree, *The Astrological School of John Abramius*

- Pseudo-Aristotle, *Traité Pseudo-Aristotélicien DU MONDE*

- Ihor Ševčenko, *The Date and Author of the So-Called Fragments of Toparcha Gothicus*

- Thomas Hyde, *Religionis Veterum Persarum*

At one point, David seemed to want to discourse endlessly on the astral magic of Harran. Harran is the city where the biblical Abraham came from (in Genesis called Paddan-Harran). Once in a while, after I got back to my office, I wrote down some of what he said from memory. There were also one or two occasions when I took along a notebook and made notes even as David spoke. But I found that this interfered with the pleasure I had in listening to him, and I didn't do it very often. Here are his words:

"I got carried away by the writings of Abu Ma'shar al Balkhi. He was from Balkh in Bactria, 786–886. Ma'shar was interested in the Hurrians. Biblically, Harran, where Abraham came from. In Harran the cult of the moon god Sin was practiced at the Temple of Sin. Incidentally, the Roman Emperor Elagabalus worshipped the moon god. There were many temples in Harran. They practiced lots of astral magic and the temples they built facilitated that."

I interrupted David just to show I was "with it."

"So this was the background when God said to Abraham, 'Get thee out of thy country, and from thy people, and from thy father's house, unto a land that I will show thee.'"

"Right. But I was talking about the temples. In Harran there was also a temple to the divine nous. Divine reason. Now that was nice, isn't it? Where today are there temples to reason, divine or otherwise? The universities? Hah!

"The Hurrians were the oddballs of the region, isolated from the nearby groups. We know that in the year 830, long, long after Abraham's residence, there was a formal meeting between the caliph and a delegation of the Hurrians.

Abu Ma'shar's manuscript on astronomy

"The caliph asked 'Are you guys pagans?'

"The Hurrians replied, 'No. We are People of the Book. And we are followers of the prophet.'

"The caliph: 'What book?'

"The Hurrians: 'The writings of Asclepius.' (Asclepius was Apollo's son, the god of medicine.)

"The caliph: 'What prophet?'

"The Hurrians: 'Hermes Trismegistus.'

"As far as I know, there haven't been any proper excavations of Harran. In the 1950s,

Rice did some digging there. Then he committed suicide. Sad. Very sad. So we have to rely on medieval Arabic descriptions of the temples. Al-Damashki gave a description of the temples of Harran. But he has very little to say. They were built along strongly geometric lines: octagons, dodecagons, icosagons. Each type was keyed to a special planet. Now there's more applied math for you. Each built out of special materials also keyed to the individual planets. For example, Saturn was lead or black stone.

Ruins in Harran

"At any rate, I got wind of the fact that around 850 Abu Ma'ashar wrote a book, now lost, on the temples of Harran. I call it the Book of Temples, and it contained, so it was suggested, descriptions of the temples, the temple rituals, and the Hurrian theory of magic. We also have allusions to a later version in Arabic, around 1401, that even contained some drawings of the temples. What a find that would be if the book could be recovered!"

"And then you went off, looking for the book?" I broke in once again.

"Yes, I went off. No luck, but when one looks, one always finds something of interest, even if it isn't what one expects."

Apollonius of Tyana
Wandering Sage and Wonder Worker
Held by Some as a Christ-Like Figure

A paragraph in David's letter to me of March 25, 1980, reads

I assume you have read, as well as Philostratus' Life and Apollonius' own Letters, the attack on the latter (or rather on both) by Eusebius. Besides these you must read the account (or reckoning) of his magical acts in Greek and Arabic given by P. Kraus in vol. 2 of his Jābir ibn Hayyan. Jābir, incidentally, should join our list of weirdoes.

Well, I never got to Jābir in P. Kraus, vol. 2, but I did get to Philostratus, so here goes.

Philostratus tells the story of how Apollonius of Tyana (first century CE) went to visit the College of the Magi in Babylon. Apollonius and the Magi were in conference for days on end, meeting every day at noon and at midnight. Damis, who was Apollonius' Boswell and Sancho Panza, asked Apollonius what he thought of the Magi. He answered, "They are certainly wise, but they are not wise on every subject." Apollonius knew he himself could give the Magi cards and spades as regards wisdom.

Apollonius of Tyana

The writings about Apollonius appear to be full of made-up stories. The one that I like best is about the lamia. Recall: poet John Keats wrote a long poem about Lamia. The lamia is all over the map folktale-wise. Here is my version; I call it "Lamia or The Lady Vanishes."

When Apollonius was staying in Corinth, he had a student by the name of Menippus Lycias. Menippus was a handsome and intelligent young man, about 25 and in top physical condition. One day, while walking on the road that led out of town, Menippus was approached by a beautiful woman. The woman grasped his hand and confessed to him that she'd seen him around and that she'd been in love with him for a long time. Her speech and manner were slightly strange. Menippus questioned her about this and she explained that she was a Phoenician. She added that she lived on the outskirts of the city.

Now Menippus was a ladies' man and despite the fact that he was taking instruction in philosophy, he liked his little piece of flesh now and then. He fell hard for the Phoenician lady.

The presumptive Phoenician lady dripped honeyed words copiously. Having named such and such a trysting place out of town, she said to Menippus,

"Come there this evening. We'll share a bottle of wine the like of which I guarantee you have never tasted. I'll sing you a song in verses you would not have thought possible. And, mind you, no one will be around. We'll share our beauty between ourselves."

[And I like to imagine that she raised an eyebrow, winked, and added "If you get what I mean."]

Menippus did, indeed, get what she meant and visited her that evening. The lady seemed to be wealthy; she was generous in all things, even to extravagance. He went to her the next night and the next and they shared much beauty. In brief, she became a habit with him.

The news got around to Apollonius, Menippus' teacher.

"You're a beautiful boy. You know that, don't you? Beautiful women chase you. But you don't realize what you're in for with this particular woman. You've

been cuddling a serpent and a serpent has been cuddling you. Are you propos-
ing to marry her?"

"Yes. Tomorrow. She loves me."

"Well," said Apollonius, "you'll see me at the feast."

The wedding was held in a great banquet hall. The place was decorated
lavishly with flowers and expensive hangings. Cooks cooked, cupbearers
bore, musicians played, and the odors of delicious perfumes and exotic
foods wafted across the vast spaces. The wedding table was laid with gold
and silver dishes and the wedding guests marveled at the magnificence of
it all.

Into this Cecil B. DeMille set strode Apollonius.

"And where is the lovely and extravagant lady in whose honor we have
come?"

"Here," said Menippus.

"And who supplied all this gear, this hall, the dishes, the decorations? You or
the lady?"

"She did, Apollonius. Like you, I am a true philosopher. Like you, my cloak
is my only possession."

"Then," said Apollonius, drawing himself up to make his big speech,
"I have the honor to say to you that all of this before us is insubstantial. It is a
mirage, a dream, an illusion. It doesn't exist. The lady is a LAMIA! She is a. . . ."

The wedding guests shrank back at these words. ("No. It can't be. You
must be mistaken!")

Apollonius went on:

"The lady is a lamia. She loves to share her beauty with beautiful boys. But
more that this: she loves human flesh. And she offers her beauty to get what
she most likes. Can't you see she's been fattening you with pleasures? Can't
you see that she's after your fresh young blood? BE GONE, FOUL DREAM
[Keats]."

The lovely bride turned deadly white. She denied the charge. She laughed
nervously and appealed to the gallery.

"Everyone knows," she said, "that philosophers speak nonsense."

Apollonius persisted. The lamia began to weep. This is the moment of
crisis. What does one do when a lovely demon weeps and melts the heart?
Apollonius pressed on. He forced a confession out of her. Gradually the scene
began to disintegrate. The gold and silver plates disappeared. The trappings,

Lamia from Topsell's bestiary of 1607

the servants, the wedding guests all disappeared. Finally, the lamia herself disappeared.

Apollonius and Menippus found themselves alone in a wasteland at the edge of the city. Apollonius had exorcised a vampire. Exorcised simply means driven out. I suppose that the lamia went on to surf the territory for her next young, handsome man.

The lamia has been a fertile source for imaginative storytellers, poets, and artists. There are dozens and dozens of imagined lamias.

A lamia by Herbert James Draper (1909)

And here is how poet John Keats imagined Lamia:

> She was a gordian shape of dazzling hue,
>
> Vermilion-spotted, golden, green, and blue;
>
> Striped like a zebra, freckled like a pard,
>
> Eyed like a peacock, and all crimson barr'd;
>
> And full of silver moons, that, as she breathed,
>
> Dissolv'd, or brighter shone, or interwreathed
>
> Their lustres with the gloomier tapestries—
>
> So rainbow-sided, touch'd with miseries,
>
> She seem'd, at once, some penanced lady elf,
>
> Some demon's mistress, or the demon's self.
>
> Upon her crest she wore a wannish fire
>
> Sprinkled with stars, like Ariadne's tiar:
>
> Her head was serpent, but ah, bitter-sweet!
>
> She had a woman's mouth with all its pearls complete:
>
> And for her eyes: what could such eyes do there
>
> But weep, and weep, that they were born so fair?
>
> As Proserpine still weeps for her Sicilian air.
>
> Her throat was serpent, but the words she spake
>
> Came, as through bubbling honey, for Love's sake,

· · · · · · · · · · ·

When I told David that I had read the story of Apollonius and the lamia he said to me,

"Good. Now read how Jesus exorcised a devil. Check it out. It's in Mark."

I checked it out. It's in Mark 3:22. It's also in Matthew 12:24 and Luke 11:15.

Charles-Benôit Hase
Historian, Byzantinist, and Forger

The letter below, written when David spent a semester at Wolfson College, Oxford, served to jog my memory about this very fine fellow.

> Dear Phil,
>
> I assume you have returned from the wilderness of exaggerated egos (et alia). We left the desolation of London for the rural charm of Oxford last Saturday. Though it is still chilly, we overlook fields, trees, the Cherwell, and habitats of squirrels, ducks, and other fowl. Almost like being in Pomfret!
>
> However, in this new location I can not easily come up with a precise reference. The forger was Karl Hase (1780–1861) — but wait, I have found the citation in Segonds' "edition" of Philoponus' Περὶ τῆς τοῦ ἀστρολάβου χρήσεως καὶ κατασκευῆς. It is: Ihor Ševčenko, "The Date and Author of the So-Called Fragments of Toparcha Gothicus," Dumbarton Oaks Papers 25, 1971, 117–188. I'm sure you'll find it most amusing, and completely vindicating your admiration and respect for historical research.
>
> We look forward to seeing you in England in May. We shall be in Oxford till the end of June. Complete the triangle.
>
> With best wishes,
>
> David

David's letters to me were always printed. Here is the transcription of the previous letter:

Wolfson College
Oxford OX2 6CUD

3 March, 1986

Dear Phil:

I assume you have returned from the wilderness of exaggerated egos (et alia). We left the desolation of London for the rural charm of Oxford last Saturday. Though it is still chilly, we overlook fields, trees, the Cherwell, and habitats of squirrels, ducks, and other fowl. Like being in Pomfret!

However, in this location I can not easily come up with a precise reference. The forger was Karl Hase (1780–1861)—but wait. I have found the citation in Segond's "edition" of Philoponus' ϖερί τής τού άστρολάβου χρήσεως καί κατεσμευής. It is Ihor Ševčenko, "The Date and Author of the So-Called Fragments of Toparcha Gothicus," Dumbarton Oakes Papers 25, 1971, 117–188. I'm sure you'll find it most amusing, and completely vindicating your admiration and respect for historical research.

We look forward to seeing you in England in May. We shall be in Oxford till the end of June. Complete the triangle.

With best wishes,
David

David had mentioned Hase to me some months before he left for Oxford. I was glad to have his letter, but I did nothing about it until after he and Isabelle returned from Oxford. We were having lunch together one day when I mentioned his letter. This jump-started a brisk flow of fact, conjecture, and scandal.

"*I don't suppose that you've heard of the* Toparcha Gothicus, *have you?*"

"*You suppose correctly,*" I replied, "*Tell me about it.*"

"*Medieval Ukrainian history. And I don't suppose you've ever heard of Karl Benedikt Hase, have you?*"

"*Your supposition is on target. Well, you mentioned him once. But tell me about him.*"

"*Great scholar. Great classicist. Great ladies' man. Also a forger. He forged several fragments of the Toparcha Gothicus. At least Ševčenko thinks so.*"

"*Who is he?*"

"*At Harvard. Department of Ukrainian Studies.*" [Now Dumbarton Oaks Professor Emeritus of Byzantine history.]

"Why do you bring up these guys?"

"Well, didn't you tell me some time ago that you were interested in the origin of Napoleon's theorem in geometry?"

"I did."

"Well, you ought to go up to Cambridge and talk to Ševčenko. There's a connection between Hase, Napoleon, and geometry. He probably knows all about it."

"What's the connection to Napoleon?

"Hase, I think—well, it's been conjectured— had an affair with Hortense de Beauharnais, wife of Louis Bonaparte, whom Napoleon made king of Holland."

Hortense de Beauharnaise, queen of Holland

"So?"

"More to the point, Hase also, for a while, tutored Hortense's two young sons. History, classics, mathematics."

"I still don't see the connection."

"Think about it a bit. Read Ševčenko's article."

Since I have discussed my conjectured theory about the relationship between Hase, his two young students, and Napoleon's theorem at some length in my *Mathematical Encounters of the Second Kind*, I will skip that aspect of the story and go directly to what I learned from Ševčenko (1922–2009)—and from David, of course.

Charles-Benôit Hase, né Karl Benedikt Hase (1780–1864), was born a German but lived most of his life in Paris. He was a very famous scholar of classical Greek and Byzantine literature. He was professor at the École Polytechnique and conservateur des manuscrits at the Bibliothèque Nationale. He did a number of brilliant textual restorations. One manuscript that he worked on, that of Johannes Lydus (Byzantine historian, c. 550 CE), had been hidden for decades in a wine barrel in a monastery, and its recovery and restoration could be worked up into a romance worthy of Sir Walter Scott or Umberto Eco. He wrote intelligently on Byzantine astrolabes, so he must have known more than a bit of mathematics.

Monsieur Hase was a man with an enormous appetite for the amatory arts and pastimes. He had a (conjectured) affair with Hortense, queen of Holland. Her husband, Louis Bonaparte, was decommissioned by a fille de joie, and a

Ihor Ševčenko

bit paranoid. He would take a candle and look for Hortense's lovers, real or imaginary, under her bed.

Hase left a very blunt and scandalous secret diary written in a kind of private Greek language in which these things are implied. The secret diary has been lost, but some of the less scandalous parts have been copied out and are available. Here is Ševčenko on one of the more scandalous passages in Hase's diary:

> An intimate encounter with a *fille-de-joie* was entered [in the diary] as συνουία σύν τη κόρη της δίόδου. For a refined technical detail in the same area of endeavor a [grammatical] dual was used: [then more Greek]. Fear of the consequences attending upon συνουία appeared, classical, enough as [more Greek], but the contraption to prevent the latter is denoted by a neologism [still more Greek].

I have suppressed most of the Greek words in Ševčenko's text, as I do not want to offend the sensibilities of my more learned readers (though, these days, what Hase has written here is pretty mild stuff indeed).

I come now to the forgery. Hase was not a counterfeiter of money or a forger of signatures on checks; he was a forger within his scholarly profession (at least modern scholarly opinion believes he was). In 1819, in an annotated history of Leo Diaconus (Byzantine historian, c. 950), Hase inserted three Greek fragments, previously unpublished, that bear upon the history of Crimea in the year 989. These fragments were so skillfully and cleverly composed, with sufficient corroborative detail providing so much verisimilitude, that their authenticity was unquestioned for 50 years. And, as they saying goes, "it takes a thief to catch a thief," for in later years, Hase, "aware of the pitfalls to which a falsifier is exposed in his work," was able to unmask Constantine Simonides (1820–1867), the notorious forger of Greek manuscripts. Even now, the case against Hase has not been secured at the 100% level of certainty.

So what is in the spurious fragments? They relate to the taking of the city of Kherson in Crimea in the year 989 by Prince Vladimir.

> The first fragment related how a party headed by the narrator [of the fragment] crossed the frozen Dnieper and traveled through the steppe in the midst of a snowstorm. The second fragment dealt with an attack launched

at the approach of winter on an area ruled by the narrator and called Klimata by some barbarians. The third fragment reported the success of the narrator in repulsing the attack and spoke of an assembly of his allies and of the narrator's journey to the ruler holding sway to the north of the Danube, and of that ruler's investing the narrator with the government of the Klimata.

Order of Merit for Arts and Sciences; won by Hase in 1849

Ševčenko's thought is that these fragments underlined the prowess and the dominance of the Russians against their enemies. This would have appealed greatly to a Count Nicholas Rumjancev, chancellor of the Russian empire, who was a considerable patron of Hase.

Thus do historians change what God is not able to change: they can alter the past. You and I, dear reader, are very likely not greatly interested in the minutiae of medieval Crimean history, but allow me to assure you that there are those who are vitally interested. David said so.

Abu Rayhan al-Biruni

F rom David's letter to me on March 25, 1980:

> *The extraordinary nature of al-Biruni can only come across by reading his many books. As starters one may turn to his* Chronology *and his* India, *both translated by Eduard C. Sachau; the translator's preface to the second, though somewhat inflated in its estimate of the man, is also good. A book that will amuse you (written by a contemporary, and undoubtedly an acquaintance of al-Biruni) is al-Tha'alibi's* Lata'if al-ma'arif, *translated by C. E. Bosworth. Therein you will learn that just four Muslims have begotten each a hundred children, and four have killed each a million men; these and other statistical and other facts await you. Do you know what geopluyy is? If not, go immediately to Bosworth.*
>
> *The microfilm reader awaits me.*
>
> *Sincerely, David*

There is a tendency of biographers to whitewash their subjects: by this I mean that mentions of eccentricities, actions, and beliefs that appear to us

Abu Rayhan al-Biruni

strange are omitted. The shorter the biographical notice, the more apt this is to occur. After all, the peculiarities of, say, the Founding Fathers of the United States are of little consequence compared to their political achievements. David, who read deeply, was able to smoke out strange and curious facts about his subjects. I, reading occasionally and more superficially, have been much less successful at it. Thus, although he assured me that Abu Rayhan al-Biruni (973–1048) was worthy of inclusion among the Ancient Loons, all I have been able to garner is very "straight stuff." And alas, David is no longer able to guide me to the "odd stuff."

Al-Biruni is one of the great polymaths of all time. He made contributions to astronomy, astrology, mathematics, physics, and medicine. He was a linguist, a traveler, and a writer of history. His name is mentioned in every decent history of mathematics. One of his striking contributions to spherical trigonometry was his solution of the "Qibla problem," that of determining in which direction a Muslim anywhere on the surface of the globe must turn to face Mecca when praying.

But to al-Tha'alibi's *Lata'if al-ma'arif* (The book of curious and entertaining information). Move over, *Guinness World Records,* move over and share the honors with this medieval work! In addition to the statistics that David mentioned in his letter to me you will find which city's earth is the best to eat.

The English expression "to eat dirt" means to submit oneself meekly to severe criticism for one's actions or opinions (synonyms: to eat crow, to eat humble pie). However, geophagy, the eating of earth in the literal sense—especially clay—has a totally different meaning and its practice is more widespread than one might think. Perhaps there is nutritional value to be found in dirt. At any rate, al-Tha'alibi, in a chapter devoted to the distinctive products of various cities, suggests that the Iranian city of Nishapur is noted for its edible dirt: "Its like is to be found nowhere else in the world."

Ringing Down the Curtain

L ooking over what I have written, I see that I have not discussed the cases
of Hermes Trismegistus, the source of all wisdom and magic; Jabir (Ge-
ber) ibn Hayyan, the alchemist; Milarespa, the Tibetan saint who performed
magic and sang 100,000 songs; or gYu-thog Yon-tan mGon-po, a practitioner
of ancient Tibetan medicine as part of Buddhist ritual. I have not yet looked
into the case of Maximus the Philosopher who, as David wrote me

> *was also rather mad. A neo-Platonizing theurge who taught the Emperor Julian
> the Apostate, and tried to avenge the overturning of his anti-Christian policies of
> his successors through magic. See: Pauly -Wissowa 14² 2563-2570 (Maximus 40)
> for a start.*
>
> *Hic Standum, David!*

I have said nothing about the book known as the *Picatrix*, the Latin edition of which David was editing at the time he first mentioned the book to
me.

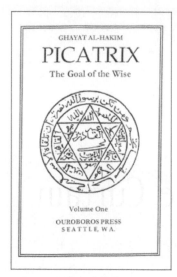

GHAYAT AL-HAKIM

PICATRIX

The Goal of the Wise

Volume One

OUROBOROS PRESS
SEATTLE, W.A.

A recent edition of the *Picatrix*

Well, what is this book? The original title in Arabic is *Ghayat al-Hakim filsihr* (Theaim of the sage) and it was supposedly written by the Andalusian mathematician al-Majriti around the year 1000. What's in it? It's a handbook of magic. What is magic? Magic (so it says) consists of three parts: the making of talismans and their effects, how the planets affect us, and the reciting of incantations.

Do you want, for example, to calm someone down who is angry with you? Then here is how the *Picatrix* recommends that you do it. When the moon has passed into the second of its 28 mansions, melt together some white wax and mastic (an aromatic resin) and use it to form the image of a crowned king. Fume it with an incense made from lignum aloes and then say "O Enedil, remove the anger between so-and-so and myself. Reconcile us." And then keep this talisman on your person.

Is this a recipe for psychiatrists or for the secretary general of the United Nations? Are their current procedures more efficacious?

Nor have I told of the times when David held me rapt with stories of the 10,000 devils of Tantric Buddhism, or the 60 times 10,000 angels mentioned in the *Sefer ha Rezin,* a book that was written by the angel Raziel and given to Noah. Yet, I believe that my readers have now achieved a good grasp of the lighter side of David, of his professional knowledge and how he provided me with many hours of entertainment.

So I now ring down the curtain, and as Scheherazade said to King Shahriah: stay tuned.

Besides, these people are all underpaid.

Further Reading

General

Acta Sanctorum.

The Dictionary of Scientific Biography.

Ivor Grattan Guinness. *The Mathematical Sciences.* Norton, 1997.

The Holy Bible.

Pauly–Wissowa (Realencyclopädie des klassischen Altertums).

Der Neue Pauly: Enzyklopädie der Antike. 2003.

Specific to the Department of History of Mathematics at Brown

John L. Berggren. "Abraham J. Sachs (1914–1983): In Memoriam." *Historia Mathematica* 11 (1984), 125.

Philip J. Davis. *The Thread: A Mathematical Yarn.* Berlin: Springer-Verlag, 1983.

Philip J. Davis. *Mathematical Encounters of the Second Kind*. Boston: Birkhäuser, 1997.

Philip J. Davis. *The Education of a Mathematician*. Natick, MA: A K Peters, Ltd., 2000.

Philip J. Davis. "Otto Neugebauer: Reminiscences and Appreciation." *American Mathematical Monthly* 101 (1994), 129–131.

Philip J. Davis. "Otto Raps My Knuckles." *Mathematical Intelligencer* 29:2 (2007), 16–17.

Otto E. Neugebauer. *The Exact Sciences in Antiquity*. Providence: Brown University Press. Numerous editions from 1951 to 1962.

Otto E. Neugebauer and Abraham J. Sachs. *Mathematical Cuneiform Texts*. New Haven: American Oriental Society, 1945.

Otto E. Neugebauer. "Abraham J. Sachs (1915–1983)." *Archiv für Orientforschung* 29/30 (1983/84), 333–334. [Includes a portrait.]

Kim Plofker and Bernard Goldstein. "In Memoriam David Pingree." *Aestimatio* 2 (2005), 71–72.

Noel Swerdlow. "Otto E. Neugebauer (26 May 1899–19 February 1990)." *Proceedings of the American Philosophical Society* 137:1 (1993), 137–165.

Gerald J. Toomer. "A.J. Sachs (1914–1983)." *Journal for the History of Astronomy* 15 (1984), 146–149.

Specific to the "Loons"

Petronius Arbiter. *Satyricon*. Amsterdam, 1669.

Elias Ashmole. *The Diary and Will of Elias Ashmole*. Oxford: Clarendon Press, 1927.

John Aubrey. *Brief Lives*. Suffolk, England: Boydell Press, 2004.

(Venerable) Bede. *The Life and Miracles of Saint Cuthbert, Bishop of Lindisfarne*. [Many editions available.]

Helena Petrovna Blavatsky. *Isis Unveiled: Collected Writings*. Wheaton, IL: Theosophical Publishing House, 1972.

W. Burkert. *Weisheit und Wissenschaft: Studien zu Pythagoras, Philolaos, und Platon*. Nürnberg: H. Carl. Revised and translated by E. L. Minor, Jr. as *Lore and Science in Ancient Pythagoreanism*. Cambridge, MA: Harvard University Press, 1972.

Charles Dodgson. "A New Theory of Parallels." In *Curiosa Mathematica* by Lewis Carroll. Oxford: Oxford University Press, 1888.

Henry Chadwick. *Priscillian of Avila: The Occult and the Charismatic in the Early Church*. Oxford: Clarendon Press, 1976.

John Dillon and Jackson Hershbell, ed. and trans. *Iamblichus: On the Pythagorean Way of Life* (*De Vita Pythagorica*). *Text, Translation, and Notes*. Atlanta: Scholars Press, 1991. [Text in Greek and English.]

Katharine R. Firth. *The Apocalyptic Tradition in Reformation Britain, 1530–1645*. Oxford: Oxford University Press, 1979.

Sir James Fraser. "The Passion of St. Christopher" [Irish passion of St. Christopher]. *Review Celtique* 34 (1913), 307–325. Collected on David Woods' St. Christopher website http://www.ucc.ie/milmart/chrsirish.html.

Edward Gibbon. *The Decline and Fall of the Roman Empire*. [many editions.]

Iamblichus. *De Communi Mathematica Scientia* [On the general principles of mathematics]. Ann Arbor, MI: University of Michigan Press, 1998.

Iamblichus. *On the Mysteries of the Egyptians, Chaldeans, and Assyrians*. Mecosta, MI: Wizards Bookshelf, 1984.

John Keats. "Lamia."

Carolyne Larrington. *Women and Writing in Medieval Europe: A Sourcebook*. London: Routledge, 1995.

[Attributed to] Majriti al-. *The Picatrix* (*Ghayat al-Hakim filsihr*) [The aim of the sage]. Edited by David Pingree. *Studies of the Warburg Institute*. London: The Warburg Institute, 1986. [Latin version cited; many editions available.]

Milarepa. *The Hundred Thousand Songs of Milarepa*. Translated and annotated by Garma C. C. Chang. Boston: Shambhala, 1998.

A. L. Rouse. *Sex and Society in Shakespeare's Age: Simon Forman the Astrologer*. New York: Scribner, 1974.

Bertrand Russell. *A History of Western Philosophy*. New York: Simon and Schuster, 1945.

Sir Walter Scott. *Kenilworth*. [Many editions available.]

Chris Skidmore. *Death and the Virgin Queen*. New York: St. Martin's Press, 2010.

Ta'alibi al-. *Lat̄a'if al-ma'̄arif* [The book of curious and entertaining infor-mation]. Translated and edited by C. E. Bosworth. Edinburgh: Edinburgh University Press, 1968.

Edward Topsell. *The History of Four-Footed Beasts and Serpents.* New York: Da Capo Press, 1967.

David G. White. *The Myths of the Dog-Man.* Chicago: University of Chicago Press, 1991.

Ian Wood. *The Merovingian Kingdoms 450–751.* Essex, England: Longman Group, 1994.